Just Fodder

The Ethics of Feeding Animals

Josh Milburn

McGill-Queen's University Press
Montreal & Kingston • London • Chicago

© McGill-Queen's University Press 2022

ISBN 978-0-2280-1135-4 (cloth)
ISBN 978-0-2280-1151-4 (paper)
ISBN 978-0-2280-1323-5 (ePDF)
ISBN 978-0-2280-1324-2 (ePUB)

Legal deposit second quarter 2022
Bibliothèque nationale du Québec

Printed in Canada on acid-free paper that is 100% ancient forest free
(100% post-consumer recycled), processed chlorine free

Library and Archives Canada Cataloguing in Publication

Title: Just fodder : the ethics of feeding animals / Josh Milburn.

Names: Milburn, Josh, author.

Description: Includes bibliographical references and index.

Identifiers: Canadiana (print) 20220156212 | Canadiana (ebook) 20220156247 |
   ISBN 9780228011514 (paper) | ISBN 9780228011354 (cloth) | ISBN 9780228013235 (ePDF) |
   ISBN 9780228013242 (ePUB)

Subjects: LCSH: Animal feeding – Moral and ethical aspects. | LCSH: Animal welfare –
   Moral and ethical aspects. | LCSH: Animal rights – Moral and ethical aspects. |
   LCSH: Human-animal relationships – Moral and ethical aspects.

Classification: LCC HV4708 .M55 2022 | DDC 179/.3 – dc23

This book was designed and typeset by Peggy & Co. Design in 11/4 Sabon.

Just Fodder

Contents

Acknowledgments vii

1 Introduction:
Animals, Food, Philosophy 3

2 Feeding Animals to Animals:
The Problem of Carnivory 21

3 Animal Family 50

4 Animal Neighbours 82

5 Animal Thieves 107

6 Animal Refugees 135

7 Animal Strangers 156

8 Conclusion 178

Notes 183

References 201

Index 215

Acknowledgments

I began thinking about problems around the ethics of feeding animals while reading for a PhD in the School of Politics, International Studies, and Philosophy at Queen's University Belfast (QUB), where I was supervised by Dave Archard and Jeremy Watkins and funded by Northern Ireland's Department of Employment and Learning. For prompting me to think about food, I thank Matteo Bonotti, a QUB colleague, and, for prompting me to think specifically about the feeding of animals, I thank Katherine Wayne, whom I met at the MANCEPT Workshops.

This book was conceived and (mostly) written during my postdoctoral fellowship at Queen's University (QU). This fellowship was funded partially by a grant from the Canadian Institute for Advanced Research. During this fellowship, I was based in the Department of Philosophy and mentored by Will Kymlicka. Both Will Kymlicka and Sue Donaldson read and commented on draft chapters and discussed ideas in this book with me on numerous occasions. I am very thankful to them, as well as the wider Department of Philosophy and APPLE research group. Ideas from the book were presented at QU on several occasions, and I thank all those who engaged with them.

Further work on the book was completed after my time at QU, first while I was teaching in the Department of Politics and International Relations at the University of York (when I presented some ideas from the book at the MANCEPT Workshops), and then while I was a British Academy Postdoctoral Fellow in the Department of Politics and International Relations at the University of Sheffield (grant number PF19/100101). I offer particular thanks to Alasdair Cochrane, my mentor at Sheffield, who was good enough to read numerous drafts while I put finishing touches on the book.

Of course, this book also draws upon the many, many conversations I have had with colleagues about food and human-animal relationships. It would be impossible for me to name everyone who deserves some credit. I have found the animal ethics and animal studies community a deeply valuable one to be a part of, and so I here thank the community as a whole, rather than any particular members. Further thanks are owed to Khadija Coxon, at McGill-Queen's University Press, who has been supportive of this project from its early days, and two anonymous reviewers, who offered extremely valuable feedback.

I would also like to thank my fiancée Becky Gray – who followed me from Lancaster to Belfast, Belfast to Kingston, and Kingston to York. She is someone who gets easily as much joy from feeding animals as I do. And thanks, too, to the various animal neighbours who have accepted my hospitality; Hollie, my excitable plant-based dog; and the assorted sanctuary animals who have helped me think about different ways we could live together.

Just Fodder

1

Introduction

Animals, Food, Philosophy

Lots of vegetarians and vegans keep dogs, cats, or other companion animals. But when it comes to the feeding of these animals,[1] they face a dilemma. The received wisdom is that responsible guardians feed dogs and cats a meaty diet. But vegetarians and vegans, at least if they're vegetarian and vegan for ethical reasons, will hesitate. Animals – perhaps animals not so different from their beloved companions – suffer and die to produce meat. And vegetarians and vegans think this is a problem; it's such a problem for them that they no longer want to support it with their diet. If it's wrong to support the meat industry to feed ourselves, is it perhaps wrong to support the meat industry to feed our companions? And what of the animals our companions might eat if left to roam? What guardian of a free-roaming cat doesn't have a story of waking up to 'gifts' of dead or dying animals? These animals, too, matter. The suffering and death that our companions impose on them, on the face of it, raise ethical questions.

But the ethical dilemmas raised by thinking about animals as eaters don't end with companions. Lots of us feed garden wildlife, like the birds for whom we leave nuts and seeds. Lots of us desperately *avoid* feeding garden wildlife, like the rodents and raccoons who would help themselves to our stores or bins if given half a chance. And the questions raised are far from a purely personal affair. Collectively, through our choices, some animals are encouraged, some are discouraged. The pigeons of Trafalgar Square are brought in by tourists feeding them; the banning of feeding, though perhaps welcomed by those who have no love for the birds, leaves pigeon stomachs empty. The growing of crops attracts 'farmland' animals who would eat those crops; harvest puts these animals in mortal peril, apparently leaving (metaphorical) blood on the hands of almost all of us. Anthropogenic climate change leaves some wild animals, who may never have even encountered humans, hungry; activists call for them to be fed, or launch initiatives to feed them, while others say we should let nature be.

This is a book about these and other practical dilemmas raised when we think about animals and food. The feeding of animals raises a host of normative puzzles. Who are we obliged to feed? Who are we *permitted* to feed? What are we allowed to feed to animals? What is the role of the state in feeding animals? How might obligations concerning the feeding of animals differ from obligations concerning the feeding of humans, and why? Regrettably, there has been a distinct absence of consideration of these questions in philosophical discourse. This is despite the fact that the questions deal with pressing real-world moral and political problems; these are questions with which individuals, collectives, and legislatures regularly grapple. Individual vegans and vegetarians worry about the ethics of feeding meat to their companions, while householders fear that their love of feeding birds may be putting the birds at risk. Almost anyone who has ever come across a free-living ('wild') animal in need has asked whether she should feed this animal. States, meanwhile, are obliged to come up with public health policies concerning human-animal interaction, and laws surrounding the marketing and manufacture of feed. Others grapple with questions in a professional capacity; volunteers or employees at animal shelters and sanctuaries, for instance, must answer questions about feeding on a day-to-day basis.

This book seeks to remedy this philosophical silence by offering a critical examination of the normative dimensions of animal diets. It offers particular attention to the many and multifarious relationships that humans have with animals, and the ways that these affect the normative obligations we have towards or concerning said animals. It seeks to change the conversation; previously, philosophers have overlooked questions about animals' food, assuming that it is just *fodder*: it is unimportant, uninteresting, and unworthy of humans, both as food for the body and as food for thought. Instead, I hope to show, we need to ask about animals' foods to come up with a clearer idea of *just* fodder: what justice requires and permits of us when it comes to feeding animals; what morality implores or lauds; what we must stop doing, and what we must start doing; in what ways our relationships with animals should change, and in what ways they can stay the same.

As these motivating questions suggest, close attention must be paid to the different *kinds* of normative obligations we possess. In some cases, our obligations concerning animals as eaters can be understood in purely negative terms – the pertinent questions concern the limitations on what we can do *to* an animal. In other cases, we will have important positive obligations. We may be obliged to help particular animals, or, in other words, to do things *for* them. In some cases, our obligations will be personal. We as individuals will have a moral duty to behave in some way, or else it would be good for us to behave in that way, or, at a minimum, there is no obligation preventing us from behaving in that way. In other cases, we *collectively* may have certain obligations, to the extent that we may need to develop or establish political institutions (or else expand the remit of current political institutions) to ensure that these obligations are met. Thus, the present work – insofar as distinctions can be drawn – should be understood as spanning the continuum between moral, social, and political philosophy, between the *right* of justice and the *good* of morality. For a full understanding of the normative significance of our relationships with animals and the eating practices of these animals, we must cast a wide conceptual net.

Animals and Ethics

This book is a work of animal ethics. *Animal ethics* is an imperfect term used to describe the normative study of human relationships with animals and the obligations we – as individuals and societies – have towards or concerning them. (*Normative* questions are questions about what we *should* do, and can be contrasted with *descriptive* questions, about what we *do* do.) It thus encompasses a range of philosophical sub-disciplines, including moral and political philosophy. It also extends beyond philosophy, touching upon law, theology, and politics, among others. Animal ethics is one strand of the wider field of *animal studies*, which incorporates multifarious examinations (in the arts, 'humanities', and social sciences) of human-animal relations, both normative and descriptive. This book also draws from the literature on food. We will address this shortly, but, for now, it's worth saying something more about animal ethics.

Though the term itself only gained prominence later, animal ethics emerged as a major scholarly concern in the latter half of the twentieth century. Earlier ethicists and political philosophers had occasionally offered asides about the status of animals – either to dismiss the thought of obligations to them[2] or to make speculative comments about how one *could* argue to defend them[3] – but few philosophers had offered serious, sustained defences of human obligations to animals.[4] This began to change in the 1970s, with the publication of *Animals, Men and Morals* (Godlovitch, Godlovitch, and Harris 1971), an edited collection put together by some of the 'Oxford Vegetarians' – a group of philosophers and others living and working around Oxford University in the 1970s (Garner and Okuleye 2020). The real sea change in the status of normative thought about animals, however, came with the publication of Peter Singer's 1975 *Animal Liberation*. Singer is an Australian philosopher who was introduced to animal issues while in Oxford, subsequently becoming a member of the Oxford Vegetarians. *Animal Liberation* condemned speciesism[5] – the privileging of the interests of one being over those of another based on the species of the beings involved – and made a case for the 'liberation' of animals from the horrors of factory farming, vivisection, and so forth.

Singer's argument is utilitarian[6] – that is, it draws upon the ethical theory of utilitarianism, which says that the right thing to do is the thing that maximizes the good (roughly, pleasure) and minimizes the bad (roughly, pain). He bases his condemnation of human use of animals upon the negative consequences that said use produces – most simply, the gross levels of suffering (primarily animal, but also human) resulting from these practices. *Animal Liberation*, as well as Singer's subsequent work developing and defending a utilitarian account of animal ethics, has become the most influential and widely discussed critique of the place of animals in contemporary societies and ethical thought. Over subsequent decades, a range of alternatives to Singer's account emerged in the philosophical literature. Of particular importance was the (re)introduction of defences of animal rights.

In contrast to utilitarians, who support the action that leads to the most good, advocates of rights say that there are certain things we cannot do to individuals, no matter the consequences. Widely credited as the most significant name in this line of thought is Tom Regan, whose 1983 *The Case for Animal Rights* offers an account of rights based upon the inherent value of all beings (including at least mammals of a year old or more) who are the 'subjects of a life'. Beginning in the 1990s, Regan's case was taken up and developed by Gary Francione and Anna Charlton, who transformed it into the philosophy of *abolitionism*, according to which veganism must be considered a moral baseline, and all uses of animals (up to and including the 'use' of animals as companions) must be considered disrespectful (Francione 2007; Francione and Charlton 2015; Francione and Garner 2011).

These two strands of twentieth-century animal ethics – the utilitarian case offered by Singer, the various rights-based cases offered by Regan and his followers – can collectively be contrasted with a range of feminist approaches that challenge the rationalistic, abstract, and emotionless methodology of the rights/utility approaches.[7] Feminist writers typically reject the absolution and mathematical formalism of Singer and Regan, instead preferring to understand the obligations that humans have to animals as many and multifarious, depending on a variety of contexts, relationships, and emotional responses. Frequently, they will endorse an alternative ethics built upon the 'care' tradition, which emphasizes personal connections.[8]

All three of these respective approaches – utilitarian, rights-based, and feminist – have their strengths and their weaknesses, and all three offer contributions to a full understanding of questions about animal diet. (It is worth adding that though all three have had much to say about the use of animals *as* food and the treatment of animals *in* the food system, they have had little to say about animals' *own* diets and food.) Singer's focus on the interests of animals themselves and on sentience (that is, the capacity to suffer, to feel, for there to be *someone home*) is a strength, but he has a difficulty in looking *beyond* suffering, and his outlook is (near) inextricably tied to his consequentialist commitments. Regan can talk about things other than suffering and rejects Singer's consequentialism; additionally, his deployment of rights-talk shifts his thinking from the realm of moral philosophy towards the realm of political philosophy.[9] The justification of rights he offers, however, raises questions. Regan's reliance on axiology (value theory) leaves rights' foundations mysterious and metaphysical (Rowlands 2009, chap. 4), while abolitionism raises its own concerns – specifically, a commitment to ending human 'use' of animals leads to bizarre and unsavoury conclusions about the extinction of domesticated animals and the end of mutually beneficial human-animal relationships (Donaldson and Kymlicka 2011; Wayne 2013b). The feminist critique has merit insofar as it is true that utilitarian and rights-based approaches to animal ethics have often been blind to the moral significance of relationships, but the alternative – an animal ethics of care – moves too far in the other direction. Consequently, an animal ethics of care is much less able to ground the foundational injustice of certain actions, whether towards animals or humans. This means that none of the three main approaches, alone and unsupplemented, serves as an appropriate foundation for an enquiry into the ethics of feeding animals.

Animals, Ethics, and Politics

Luckily, this is not the end of the story. Recently, new approaches to animal questions have offered a chance to combine the respective strengths of the three main traditions. In the twenty-first century, there has been a *political turn* in animal ethics, with an emergence of an array of works addressing animal ethics but influenced more by political philosophy and political science than by moral philosophy and value theory. This new literature is distinguished and unified by the way that its contributions focus on questions of *justice*, developing new and creative ways that our states, societies, and institutions can be transformed to make the world a more just place for animals, nonhuman and human alike (Cochrane, Garner, and O'Sullivan 2018).

Prominent in this literature – one commentator characterizes it as a close-to-defining feature (Milligan 2015) – has been a focus on interest-based rights. In contrast to Regan's value-based rights, interest-based rights are justified by the fact that animals have *interests* (that is, things can go better or worse for them, they have a welfare). Not only do the interests of animals justify the existence of these rights, but they determine the rights' content. That is, it is an animal's very strong interest in a thing (say, in not suffering) that determines that they have a right to that thing, rather than to another thing. Though clearest and most developed in the work of Alasdair Cochrane (2012), interest rights form the normative underpinning of much of the theorizing in the political turn. As will be explored shortly, the present book follows the lead of this scholarship.

In some senses, it is surprising that interest-based rights approaches have not played much of a part in animal ethics until now.[10] To illustrate why, it is worth identifying two key differences between interest rights and *will rights* – the latter of which are generally taken to be interest rights' key competitor in the *human* rights literature. First, interest rights protect the interests of the rights' subject, while will rights protect their freedom of choice. So, though either approach could ground (say) a cow's right against being killed to feed to another animal, the interest right would do so because my killing of the cow is *bad for the cow*. The will right, on the other hand, would not protect the cow from my violence as such; it would protect her *right to choose* whether she is killed to be fed to another animal. Second, while will rights require beings to have a high capacity for choice, interest rights merely require beings to possess interests. Given traditional (though contestable) assumptions about animal autonomy – and so assumptions about their ability to choose – it might seem obvious that animal rights accounts should be concerned with interest rights. We have a clear idea of what it means for a cow to have an interest in not being killed so that her body can wind up in another animal's mouth. We might lack a comparatively clear idea of what it would mean for a cow to give or withhold permission to be killed for another animal.

Further, given the influence of Singer's utilitarian account, interest-based rights might seem like a natural choice. Interest-based rights are easily able to take account of the moral significance of both sentience (as sentience is sufficient, and, perhaps, necessary for a being to possess interests) and suffering (as all beings – *prima facie* and *ceteris paribus* – have an interest in not suffering), but they do not bring with them the more surprising and unpalatable consequentialist conclusions of Singer's utilitarianism. Interest rights, especially as conceptualized in the political turn, also bring with them the advantages of rights accounts; they can condemn particular modes of treatment and ground the importance of the individual. Perhaps most importantly, rights tie closely to questions of *justice* and the institutionalization of the protection of animals (Cochrane 2020). Animal abuse – whatever it may entail – is shifted from the realm of moral disagreement, which must be tolerated in the liberal state, to a foundational issue of justice (Cochrane 2010a, chap. 8; Garner 2013, chap. 3; Smith 2012, introduction). Work in the political turn, then, gives us the opportunity to step beyond saying that it would be good, or nice, or virtuous to (for example) protect garden birds from our cats, to saying that it is appropriate for the state to require us to protect garden birds from our cats. But, what's more, it gives us the chance to step beyond questions about our individual obligations, and think about wider structural questions. For example, I can do little, alone, to change how food is grown, and thus little to change our society's relationship to the animals who are attracted to and want to eat farmers' crops. Working with others, thinking through a political lens, more can be done.

Accounts of animal ethics drawing on interest-based rights – especially when influenced by political philosophy as well as moral philosophy – provide the opportunity for a valuable coming-together of utilitarian and rights-based accounts of our obligations to animals, and so offer a genuine chance of a new consensus in animal ethics. For this reason, an interest-based rights approach to animal ethics provides a suitable foundation for an enquiry into the ethics of feeding animals. (Though it is worth stressing that, while interest-based rights will be touched upon at several points throughout the work, many of the arguments are comprehensible independently of a commitment to interest-based rights. Little of what I say stands or falls on the acceptance of interest-based rights.)

There is, however, a problem. While an account of interest-based rights is a useful starting point for understanding the rightness of wrongness of our treatment of animals in terms of the *negative* obligations we have towards them – and so for answering questions about whether it is wrong for us to inflict suffering on animals, or kill them, or imprison them, or destroy their habitats, or similar – it has some difficulty in taking account of the positive duties we may or may not owe to them, like duties to feed. Indeed, animal ethics traditionally has been reluctant to pay much attention to positive obligations we have towards animals. This is because – intuitively, and, I hold, in fact – the positive obligations we have towards animals depend, in a very large way, on our relationships with them (Palmer 2010).

If this sounds controversial, consider the following: in 2019, I adopted a dog, Hollie. Clearly, I have certain positive obligations towards Hollie that I do not have towards a dog that a stranger, a neighbour, or even a friend has adopted. Crucially, for example, I have an obligation to feed her. Also, while I may have some obligations to wild animals – it's wrong for me to capture and torture hares – I will not have all the same obligations towards them that I have towards Hollie. Lines, of course, blur. If I had a young daughter, and she adopted a dog, I would clearly have certain kinds of positive obligations towards that dog. And what of the 'wild' animals with whom I enter some kind of relationship, such as the sparrows I feed in my garden? Or the raccoons who – when I lived in Canada – tried to rustle through my bins? These kinds of questions have motivated some of the feminist critiques of the more traditional approaches to animal ethics, and are precisely the kinds of puzzles with which I grapple in the present work. For this reason, it is necessary for this book to step beyond the interest-based rights approach to animal ethics and to seriously explore the normative significance of the relationships we have with different animals.

It is important to note at this juncture that we would be wrong to imagine that all of the pertinent relationships are personal ones – the personal relationship I have with my dog, or your dog, or the sparrows I feed, or the local raccoons. The normatively relevant relationships in some cases might be best conceptualized as *political* relationships. Some animals are a part of our state/society, insofar as they contribute to it and are affected by political decisions made. So, for example, some animals might seem to be not just family members or friends, but co-citizens (Donaldson and Kymlicka 2011) or workers/colleagues (Blattner, Coulter, and Kymlicka 2020). In trying to understand the many and multifarious relationships we have with animals – and, importantly, the normative significance of them – we must keep one foot firmly in the realm of the political, even if the other stays in the moral.

In recent years, there have been two highly important works of animal ethics that focus upon contextual or relationship-based obligations to animals. First, Clare Palmer's *Animal Ethics in Context* (2010) explores the moral dimensions of human relationships with animals, developing a relational approach according to which we have little obligation to aid free-living animals unless we have impacted them in some way, even if we retain considerable obligations towards our companions. Second, Sue Donaldson and Will Kymlicka's *Zoopolis* (2011) explores the variety of political relationships humans have with animals, splitting them into three discrete groups based on the relationship they have with the zoopolis (i.e., the mixed human-animal state). Animals who are members of the zoopolis ('domestic animals', including companions, animals currently used on farms, and so forth) are offered citizenship, with the various rights and responsibilities that this entails. Animals who live in and among, or are otherwise dependent upon, the zoopolis without being a part of it ('liminal animals', including wild animals adapted to living in human-made niches) are offered denizenship, much as are human refugees, human non-citizen workers, and similar.[11] Animals who live separate from the zoopolis, in their own spaces and territories, are offered *sovereignty*.

Both Palmer on the one hand and Donaldson and Kymlicka on the other recognize, despite their focus on relationships, the importance of more basic negative and universal obligations towards animals. Palmer does this through an acknowledgment of the importance (if incompleteness) of 'capacity-oriented' approaches to animal ethics (2010, chap. 2), while Donaldson and Kymlicka do so by endorsing 'Universal Basic Rights for Animals' (2011, chap. 2). Both draw upon the importance of animal interests when it comes to understanding these things. This provides a compelling model for the exploration of the ethics and politics of feeding animals; although we must begin with a recognition of the negative obligations owed to animals (as derived from the interest-based rights approach), we must then move on to probe the significance of relationships.

Though *Animal Ethics in Context* and *Zoopolis* will be important influences, and certain aspects of them will be explored in greater detail in subsequent chapters, my exploration will expand upon and diverge from the ideas presented in those works in several important ways. While Palmer's work draws primarily from moral philosophy and Donaldson and Kymlicka's is firmly in the realm of politics, the present work will draw from both fields, moving between them as the cases and issues dictate. Indeed, the cases and issues are what most clearly distinguish this work from where academic animal ethics has gone before. In its focus on the issue of food and especially animals as eaters, the present work addresses issues previously overlooked.

Food and Philosophy

Given the present work's focus on diet, it is appropriate to say something about philosophical work on food, as this is the second broad area from which the book draws. Though, to repeat, this is a work of animal ethics, the philosophy of food can offer important insights and arguments to the present enquiry. As such, it is worth saying something about what I understand by the term *philosophy of food* before beginning to consider animal diet in earnest. Parallel to their treatment in animal ethics, questions related to animals' food and diet have been almost entirely absent from philosophical analysis focussed on food.[12] This means that this literature, too, is importantly incomplete.

Unlike animal ethics, which enjoys a relatively stable place in the philosophical canon of the twenty-first century, contemporary academic philosophy has an ambiguous relationship with food scholarship. Nonetheless, discussions about food have had a place in the philosophical literature for some time. Western philosophy has been idiomatically described as a series of footnotes to Plato, and the philosophical exploration of food might be characterized as simply one group of these footnotes (Kaplan 2012, 1). After all, Plato discussed food in a variety of contexts.[13] He was not alone in his consideration of these kinds of questions; in Ancient Greek philosophy, philosophers discussed food as much as sex and medicine (Zwart 2000), two topics undoubtedly part of the philosophical mainstream today. As such, food is not something that could be considered absent from the history of philosophy. What *has* been absent, however, is the philosophical study of food as a topic worthy in and of itself. For this, we must jump forward to the twentieth century.

Food studies is a name for the interdisciplinary academic examination of food and food-related issues, and is, accordingly, very broad. *Food ethics* is much closer to present concerns. The discipline can be defined simply as the 'deliberative inquiry into the normative dimensions – the reasons and rationales – of food' (Thompson 2016, 67). As observed by David Kaplan (2012), philosophers in the twentieth century who considered food-related issues would sometimes describe themselves as engaging in food ethics, seeing their work as being a particular instance of applied ethics. That said, food ethics, both the term and the sub-discipline, is not restricted to the twentieth century, and may indeed be on the rise. 2016, for example, saw the launch of a new journal dedicated to the area, simply entitled *Food Ethics*.

As with animal ethics, moral perspectives have dominated discussions of the normative aspects of food. This has led some theorists to call for – and begin to theorize – a *political philosophy of food* (Ceva and Bonotti 2015).[14] Unlike food ethics very narrowly construed, a political philosophy of food deals with questions of justice raised by food, as well as with the institutions that govern and distribute food. For instance, rather than asking about whether we have obligations not to be obese, political philosophers of food might ask whether the obese (perhaps including obese animals) are the victims of injustice, and ask what state institutions should be doing about this – closely informed, no doubt, by the social scientific research which expands on what state institutions *are* (not) doing about this. Though Emanuela Ceva and Matteo Bonotti are right that normative consideration of food has been problematically limited to moral questions in many analyses, there is little value in separating the political philosophy of food from food ethics. Instead, we should follow Kaiser and Algers (2016) in classifying political questions about food as a subset of the broader field of food ethics, just as the work in the political turn in animal ethics is still – as the name suggests – part of animal ethics *simpliciter*.[15] Indeed, Hub Zwart (2000), in a survey of the history of food ethics up to the end of the twentieth century, identifies a recognition of the importance of social (and, we might add, political) elements and factors as a defining part of contemporary food ethics, and part of what separates it from the food ethics of the ancient, medieval, and early modern (Western) worlds.[16]

Food ethics, even understood broadly to include the political philosophy of food, still does not capture all that philosophy can offer to thinking about food – including in the context of thinking through the normative dimensions of animal diets and food. Importantly, examining the food of animals requires thinking about food beyond thinking about ethics. For Kaplan, scholars examining food often 'sell themselves short' by describing their work as 'food ethics'. 'Philosophers', he notes, 'do more than treat food as a branch of ethical theory. They also examine how it relates to the fundamental areas of philosophical inquiry: metaphysics, epistemology, aesthetics, political theory, and, of course, ethics' (2012, 1). Philosophers exploring food, even if their focus is on ethics, move between moral philosophy, political philosophy, aesthetics, metaphysics, and other branches of philosophy. This is not a problem if done carefully and reflectively, and, indeed, it can significantly improve the rigor and reach of these discussions.[17]

We thus arrive at an appropriate characterization of the literature from which I will be borrowing here: *the philosophy of food*. Kaplan acknowledges that the term sounds somewhat novel; this, in part, might be because it 'is much easier to treat food as a mere case study of applied ethics than to analyze it as something that poses unique philosophical challenges' (2012, 2). The present work will not engage substantively with many philosophical areas with which philosophers of food are concerned – for example, aesthetics, the philosophy of perception, and phenomenology will not be referred to at any length (see Korsmeyer 2002) – but it nonetheless seeks to do more than use food as an ethical case study. Instead, it will probe the significance of food and food-related relationships themselves. As such, I engage with the philosophy of food as Kaplan understands it, and not *mere* food ethics. While my focus will overwhelmingly be on the moral and political philosophy of food, puzzles about the fundamental nature and meaning of food will be ever-present. This work is perhaps best characterized, rather than as animal ethics taking inspiration from food ethics, as a work of animal ethics drawing insight from, and contributing to, the philosophy of food.

A New Direction for Animal Ethics and the Philosophy of Food

Animal ethicists, both in moral and political philosophy, have always had a great deal to say about using animals for food. Condemnations of (factory) farming, praise for veg(etari)anism, and a dismantling of arguments in defence of meat-eating are found in most classic texts in animal ethics. This much is well known. Perhaps for this reason, works in the philosophy of food (at least when a moral/political component is present) almost invariably address the issue, frequently with contributions from animal ethicists/'pro-animal' thinkers. For example, Elizabeth Telfer, one of the first to seriously write about the philosophy of food, cites concern with the ethics of meat-eating as one of her key motivations (1996, ix), and offers a condemnation of meat-eating and defence of vegetarianism in her *Food for Thought* (1996). The fact that animal ethics finds a place in the philosophy of food is not surprising, for at least three reasons. First, there is the already-mentioned focus on food in the animal ethics literature. Second, animal ethics has helped spread concern for food ethics, both in an academic and a popular context (Thompson 2016, 63). Third, and most obviously, ethical questions about animals make up a substantial part of the remit of food ethics/the philosophy of food (see Kaiser and Algers 2016; Kaplan 2012).

Sadly, animal ethicists do not always engage with the philosophy of food, to the detriment of the sub-discipline. Thus, despite my worries about some aspects of the philosophy of food, I do hope that animal ethicists can be ready to learn from the philosophy of food when appropriate. It is my claim that animal ethicists have occasionally made uncritical assumptions about food, and, indeed, that there may be ideas and resources in the philosophy of food that animal ethicists can use. As a single concrete example, consider the following. In chapter 4 of this book, I will discuss the ethical puzzles raised by the common practice of feeding garden wildlife. If we think about this as a puzzle for the philosophy of food, new avenues present themselves. Philosophers of food are frequently keen to unravel the kinds of relationships and obligations that arise through our feeding of others. For instance, some have made efforts to resurrect the 'lost' virtue of hospitality (e.g., Boisvert 2014b, chap. 2;

Boisvert and Heldke 2016, chap. 1; Telfer 1996, chap. 5), and this may offer a way to understand (or contextualize) the relationship involved in feeding these animals. Now, this is not to say that philosophers of food have all the answers; it may be that these ideas cannot be extended to include human-animal relations, or perhaps that they are problematic. But animal ethicists cannot know this unless they are willing to explore and engage with work going on in the philosophy of food.

This brings us back, then, to the focus of this book. Rather than following in the footsteps of animal ethicists and philosophers of food who have gone before by exploring animal ethics and food through an examination of human diet, my focus is on animals as eaters, and the various questions this raises. This focus will not be exclusive; discussion of human diet will sometimes enter, or else what I say about animal diets may have certain consequences for human diets, too. As well as bringing together animal ethics and the philosophy of food, my focus will allow a chance to revisit (through a new lens) certain puzzles in animal ethics, as well as to shine a light on interesting normative issues that have been overlooked – in both animal ethics and the philosophy of food. This exploration will offer solutions to practical problems, and draw together animal ethics and the philosophy of food, to the benefit of both sub-disciplines.

Outline of the Book

Let us turn to the structure of the present work. After an initial exploration of a central recurring issue – that many animals eat meat – the book will be arranged around the different kinds of relationships that we have with animals. It will move from those animals 'closest' to us to those furthest away.

Chapter 2 addresses the problem of carnivory. This is a problem that arises in questions about the feeding of animals that does not arise in questions about the feeding of humans. In short, it is the problem that many animals are carnivorous – they have evolved to rely primarily or entirely upon the consumption of animal protein. Thus, their meat-eating is a problem of biology, not of (as in the case of humans) ideology. An ethical dilemma thus arises. How can we balance the (ostensive) 'need' for meat possessed by the carnivores we are feeding with the interests of their would-be food to not be slaughtered? The chapter canvasses a range of possible solutions, arguing that developments in our understanding of veterinary nutrition may offer the prospect of plant-based diets for (even) carnivores, while a range of potentially harm-free sources of meat are endorsed for those animals who (for whatever reason) cannot thrive on plant-based food. The conclusions of this chapter mean that animals we feed, no matter their biology or our relationship with them, can be nourished without the need for rights-violating harm to other animals.

Chapter 3 explores our relationship with our 'animal family' – companion animals, or pets. It may seem obvious that we have an obligation to feed our companions, but getting to the bottom of *why* will provide important normative foundations for the ethics of our relationships with animals more broadly. Retaining a focus on meat-feeding, the chapter explores a range of factors that we may think should motivate our choices about how to feed our companions. It then looks to relationships beyond the guardian/companion bond, ultimately arguing that companions' *right* to be fed speaks to the need for their voices to be heard in political decision-making fora.

Chapter 4 moves on to consider our animal neighbours; those urban and suburban animals who live among us, without being part of our household. Some of these animals we choose to feed, like garden birds. Some, however, we go out of our way to *avoid* feeding, like the rats attracted to under-maintained compost heaps. It is proposed that the virtue of *hospitableness* provides a valuable lens through which to consider our relationships with these animals. In short, we can choose to extend hospitality to our neighbours, but we are under no obligation to do so. When we do extend hospitality, we take on certain obligations concerning our guests' wellbeing.

Chapter 5 considers a particular kind of animal neighbour who has created some normative problems for animal ethicists: those animal 'thieves' who are attracted to our farmland, because they want to eat the crops we grow for ourselves. These animals are put at mortal risk by arable agriculture, and this provides an intriguing challenge to veganism. It is argued that the existence of these animal thieves actually calls for the development of vertical, indoor agriculture. This is a form of arable farming that could, in principle, produce food without *any* impact on our animal neighbours.

Chapter 6 turns to the animal refugees – wild animals in need of aid. In particular, it explores the institution of the wildlife rehabilitation centre. It is argued that not only should animals in these institutions not be fed meat, but that 'rehabilitated' predatory animals should not be released. This is because the involvement of rescuers in these animals' lives places significant responsibility for the animals' subsequent actions on the heads of humans. Released predatory animals will, in all likelihood, put significant (metaphorical) blood on the hands of rescuers – respect for the rights of the animals that these predators would kill and eat means that they should not be released.

Chapter 7 looks to animal strangers. The emergence of scholarly literature defending intervening in nature for wild animals' own sake encourages us to rethink common-sense assumptions about what we do and do not owe to wild animals. It is argued, however, that even if we have a duty of beneficence to protect wild animals from their predators, there are limits to what we may permissibly do in pursuit of that goal. Thus, for example – absent a clear proposal for how this could be achieved respectfully – we should stop well short of reordering ecosystems. On the other hand, it is argued that even if we do *not* have an obligation to feed wild animals in need, we can gain such an obligation when we become entangled with these animals in a morally salient way. Our responsibility for harms occurring to animals as a result of anthropogenic climate change may provide an example of just such a morally salient entanglement.

Chapter 8 concludes with a list of take-home conclusions from this book's analysis. As I will acknowledge alongside these conclusions, I do not pretend that this book has captured all the relationships that we have with animals. In particular, it has not investigated the contours of the most important food-related relationship we have with animals: the fact that we farm them for our own food. Though the feeding of these animals raises a whole host of ethical and political questions, no chapter is devoted to this relationship for a straightforward reason. As will be explored in chapter 2, this is a relationship that should not exist. While there are undoubtedly important considerations for activists relating to the feeding of these animals, this is not a book about activism. It is a book about how we should relate to animals – I am going to offer no space to how we should conduct relationships that should not exist at all.[18]

As well as allowing a chance to answer questions that have been all but ignored in animal ethics and the philosophy of food, the lens of animal diet offers us a new chance to explore philosophical questions about our own diet, and, perhaps more significantly, offers us an alternative way to think about animal ethics more broadly. Indeed, as the different food obligations we have towards and concerning particular animals are explored and explicated, a novel account of animal ethics will take shape: one which is sensitive to the varied relationships we have with different animals while recognizing that all animals are owed a certain baseline level of respect. It will be argued that the extent to which we are obliged to aid animals – both as individuals and as societies – depends upon our relationship with them and human involvement in their lives. Thus, for example, our positive obligations are far greater towards companion animals than wild animals, but we can gain obligations towards wild animals (even quite significant obligations) by interfering with them in various ways. Why this is the case, what this means in practice, and the significance of various ways that humans can be involved with or relate to animals will be explored as the book develops.

Before we can explore in earnest the different relationships that we have with animals and what this might mean for the food obligations we have towards them, we must tackle one of the most practical and vexing issues raised when it comes to feeding animals, and one curiously overlooked in academic animal ethics (Milburn 2017b): the question of feeding meat to other animals. As will be seen, though there is an analogous and much-discussed normative puzzle when it comes to *human* diets, the ethical issues at stake with animals are importantly different. This question is an appropriate place to start not only because it is a practical puzzle faced by a great many, but because questions about the feeding of meat to animals will recur throughout any systematic study of the ethics and politics of feeding animals. Let us, then, begin.

2

Feeding Animals to Animals

The Problem of Carnivory

Animal lovers caring for other animals face a moral dilemma when it comes to feeding. Many animals under human care – including cat and dog companions, and all kinds of carnivorous and omnivorous animals in other contexts – are typically fed diets rich in animal protein. But this seems to conflict with our duties to the animals killed to produce this animal protein. Why should the eat*en* be killed for the sake of the eat*er*? Hank Rothgerber (2013; 2014), a social psychologist, has coined the term *vegetarian's dilemma* to refer to the choice faced by those puzzling over whether to feed meat to their companion animals. I have elsewhere (Milburn 2017b) argued that there is an *animal lovers' paradox*: given that many animal lovers feed meat to the animals they care for, it seems that, paradoxically, they could reduce the amount of harm they are doing to animals by *stopping* being an animal lover.

The option of feeding omnivorous or carnivorous animals a plant-based diet frequently elicits incredulous responses and even charges of animal cruelty. These charges, however, are curiously selective; it seems that some people who are deeply concerned about putative 'cruelty' to animal eaters are decidedly *un*concerned about the death and suffering of the animals eaten (Gillen 2003, 45). Indeed, we can see this selectivity played out in vivid terms when it comes to the question of what goes into pet food.[1] (Purpose-made pet food, of course, represents only some of what is fed to companion animals, and companion animals represent only a small proportion of the animals that people do or could feed.) The food studies scholars Marion Nestle and Malden C. Nesheim, despite their apparent skepticism concerning both veg(etari)an pet foods (2010, 229–34) and animal rights activism (2010, chap. 25), express deep concern, even horror, at the prospect that pet food may contain the bodies of dogs and cats (2010, 86–9) – even while lauding the 'public service' that the rendering and pet food industries do in finding uses for the bodies of animals *other* than cats and dogs (2010, 82, 304). The radically different ways that these authors think about obligations to members of favoured species and members of disfavoured species[2] is reflective of a common view, and warrants challenge.

One way to offer this challenge is to advocate plant-based feeding. Despite the hostility sometimes directed at those who feed carnivorous or omnivorous animals plant-based diets, the possibility is now more convenient than ever – at least for companion dogs and cats. Vegetarian or fully plant-based pet food is becoming increasingly available, and the pet food industry is beginning to ask questions about the use of animal products (Knight and Leitsberger 2016; Ward, Oven, and Bevencourt 2020). For other companion animals – and non-companion animals we feed – things may be a little trickier.

Whether it is permissible to feed meat to animals is one of the most important ethical quandaries raised by a focus on animal diets. The dilemma – whether faced by guardians of dogs or cats, householders setting up bird feeders in their gardens, would-be sanctuary operators, or activists aiming to help wild animals in need – is the practical face of a philosophical puzzle that I call the problem of carnivory. *If* the animals we are feeding are carnivores, then familiar arguments about meat-eating among humans cannot be simply expanded to include them. This is because different reasons for meat-eating are at play. Simply put, rather than consuming meat because of ideology, these animals consume meat because of physiology. More problematic still, and again unlike humans, these animals (putatively) need to eat meat to survive, or, at least, thrive.

These puzzles will be explored in the present chapter. It is essential that the question is addressed at this early stage of the book, not only because it is a real ethical quandary faced by activists, vegans, and animal lovers – as well as a challenge levelled at them by skeptics ('But how will we feed cats?') – but because carnivory will be a recurring concern as we explore different kinds of human-animal relationships. First, I will present the ethical background against which my exploration of the problem of carnivory will take place; that is, the moral and political case for veganism. However, it is worth stressing that the problem of carnivory is far from a problem *solely* for vegan positions. Indeed, I will later suggest that the ethical question of whether we should be feeding meat to animals may be pressing for some who are unconcerned with *human* consumption of meat. Second, I will identify precisely what the problem of carnivory entails, and why it presents a distinct puzzle for animal ethicists. Finally, I will address several potential solutions to the problem of carnivory: Sticking with business as usual; Eliminating carnivorous animals; Switching to plant-based diets; Sourcing animal protein in surprising ways (scavenged meat, eggs, cellular agriculture, and non-sentient animals). Through this exploration, we will see both potential individual, moral solutions and societal, political solutions.

Veganism for Humans

A great many books in animal ethics, including many of the discipline's classic texts, give over considerable space to arguing in support of (human) veganism.[3] A few books in the philosophy of food do the same. However, this book will not follow this lead. Instead, I will mostly be taking it for granted that humans are obliged to be vegan so that I can move on to examine the diets of animals. Nonetheless, given that my claims about human veganism will seem surprising to some readers, it requires a short, necessarily incomplete, defence.

A simple way to ground the moral necessity of veganism begins with the observation that the production of animal-derived foodstuffs necessitates the death and suffering of many animals. Somewhere upwards of 60 billion farmed terrestrial vertebrates are killed every year for food. This does not include terrestrial animals who are killed or die in pursuit of food, such as diseased animals, nor does it include terrestrial animals who are hunted for food. Most significantly of all, it does not include the far greater number of *marine* animals who are killed every year for food. Marine animals are counted by weight rather than by quantity, meaning that the number of deaths is not known.

The overwhelming majority of farmed animals live short lives full of pain and suffering – I note pain *and* suffering because these animals frequently face considerable non-physical suffering, caused by, for instance, boredom, frustration, separation from family, and the inability to engage in activities that are important to them. They are bred to grow in ways beneficial for those who farm them, and problematic for the animals themselves. For example, 'broiler' chickens will grow so quickly that their bones will snap under the weight of their bodies. Various money-saving practices can be very harmful for animals. Highly painful procedures, such as dehorning, debeaking, and castration, will be carried out without anaesthetic; animals will be forcibly impregnated; and animals will be kept and transported in cramped, bare conditions. Some more expensive animal products are produced without some of these practices – the precise procedures vary between different 'humane' products – but almost all farmed animals turned into meat face the same grisly end in a slaughterhouse, invariably at a fraction of their 'natural' lifespan.

Egg and milk production do not fare any better. Animals kept for these reasons face comparable conditions to animals farmed for meat, and are also bred to have certain profit-maximizing traits, regardless of the effect on their welfare. In addition, animals kept for milk and eggs face slaughter/disposal at the end of their optimally productive life, again at a fraction of their 'natural' lifespan. These industries also, at present,[4] have the 'problem' of unwanted young. Only female chickens lay eggs, and so male chicks born in hatcheries are killed within hours of birth; a variety of methods are used, including grinding and suffocation. (Male chicks are of little use to the meat industry, as different lineages of chickens are used to produce flesh – though chicks are sometimes utilized as food for animals.) 'Dairy cows' must be continually impregnated to produce milk. The methods used to impregnate these animals, the immediate or near-immediate separation of cow and calf, and the veal industry – which is where male calves are typically utilized – all contribute to suffering and death.

Animals matter morally. No grand proclamation of equality between humans and animals is required to endorse this claim; only a recognition that some of the things we could do to an animal are wrong because of what we owe to the animal herself. Consequently, we can inflict suffering upon, or kill, animals only when there is some reason of comparative significance justifying our action. Given the suffering and death necessitated by our dietary practices, we must ask if there is such a reason in this case. We do not need to eat animal products for our health – veganism is a perfectly viable option for humans at any stage of their life (Craig and Mangels 2009). For a great many of us, our choice to consume animal products – if we do – ultimately comes down to preference and convenience. But marginal increases in our pleasure and our convenience sound like weak reasons for inflicting suffering upon and killing morally significant beings.

Gary Francione and Anna Charlton (2016; cf. 2013) ask us to imagine Fred, a man who keeps animals in his house to hurt and kill but who is otherwise morally upstanding. Imagine we were to confront Fred about his torture of animals. He might justify his choice by saying that he enjoys torturing these animals. If we implored him to seek enjoyment elsewhere, he might allow that there are other ways he could enjoy himself. Perhaps he enjoys watching television and playing video games, but feels that buying a television set and a games console, and then working out how to connect them, just sounds like too much trouble. We would not find Fred's argument convincing, and so we should not find convincing the argument of the defiant non-vegan who finds animal products too enjoyable and convenient to give up – providing, of course, we agree that animals matter morally.

This argument gives us a reason to believe that eating animal products is immoral, and a clear explanation of why we as individuals should be vegans. Of course, there are replies that could be offered. For example, in this book's introduction, I noted that animal ethicists have sometimes been too quick to downplay the value of food, and perhaps that is what I am doing here. Perhaps there is something *more* than mere pleasure or convenience in the offing in some cases of human meat-eating. Perhaps, for instance, the consumption of animals (or their hunting) has intense cultural significance, and this significance is not reducible to mere pleasure or convenience (cf. Ciocchetti 2012). Or might there be people with unusual health concerns making veganism comparatively difficult? Might there be people living in places where access to plant-based foods is limited? Might there be *some* level of lost taste experience that, in principle, justifies harming animals (Kazez 2018)? Perhaps.

And a full exploration of the ethics of veganism needs to address these questions. But this is not the place for a full exploration.

Even if this argument gives us good moral reason to be opposed to the eating of animal products, it does not give us a straightforward reason to believe that eating animal products is *unjust*, and should be prevented by the state. Strictly speaking, I do not hold that the eating of animal products is unjust (see, e.g., Milburn 2020b). I do hold, however, that standard practices in farming and fishing – including, most obviously, the killing of sentient animals – are unjust, meaning that I believe that animal agriculture should be (all but) prevented by the state. My argument for this draws upon the interest-based rights approach to animal ethics.

The most fundamental interest possessed by animals is an interest in not suffering. All sentient animals have this interest; any being capable of suffering has – all else being equal – her wellbeing set back when she suffers. This interest is strong enough for animals to possess a *prima facie* right against having suffering inflicted upon them. (The significance of this being a *prima facie* right, as opposed to some other kind of right, will be explained shortly.) The second most fundamental interest possessed by sentient animals is the interest in continued life. Continued life is in the being's interest insofar as the chance of any positive experiences are in the being's interests. Further, many animals possess cognitive complexity sufficient for them to be able to conceive of themselves as beings existing over time, and/or have ongoing projects that they wish to complete. These strengthen the existing interest in continued life possessed by such animals. This interest in continued life grounds, for those animals who possess it, a *prima facie* right against being killed.[5] (Other right-grounding interests may be thwarted by animal agriculture, but let these two suffice for present purposes.)

The *prima facie* rights possessed by animals against being killed and against having suffering inflicted upon them are violated on an enormous scale by the institutions of animal agriculture, hunting, and fishing. But violating *prima facie* rights is not unjust; instead, it is the violation of concrete rights that is unjust. If these animals' *prima facie* rights convert, in these cases, to concrete rights, then these animals are the victims of considerable injustice. As with all such rights, these *prima facie* rights convert to concrete rights if there are no equally pressing countervailing interests the realization of which are incompatible with these rights being concrete. It is undeniable that humans have very strong interests in having access to plentiful and nutritious food, in making a living, and in having pleasurable (food) experiences. And it is undeniable that, at present, many people realize these interests precisely because of animal agriculture (or hunting, fishing, etc.). However, this is not necessarily so. Humans could have access to plentiful, nutritious food without animal products. Humans could make a living without having to kill animals and make them suffer. And humans can have pleasurable experiences, including food experiences, without the use of animal products. Consequently, these rights do translate to concrete rights, and animal agriculture – for human consumption, at least – is unjust.

The preceding – drawing from Alasdair Cochrane (2012; see also Garner 2013) – is not meant to be a complete and irrefutable argument against animal agriculture; instead, it sets out some of the presumptions with which I will be working in the following analysis.

So far in this chapter, I have provided an outline of the reasons for which I hold that veganism is both a moral requirement and a requirement of justice. The arguments deployed can allow that non-veganism is acceptable (if regrettable) in cases in which veganism is genuinely unfeasible. However, for the most part, these arguments provide, in my view, compelling justifications for the immorality and injustice of non-vegan diets in the West. Neither argument relies explicitly upon the problematic nature of animal products in and of themselves (though other arguments might). Instead, they point to the problematic aspects of animal agriculture – aspects that remain whether we are raising and killing animals for human food or for the food of other animals. Either way, humans are inflicting suffering upon and killing animals, and, either way, those who purchase these products are financially supporting animal agriculture.

These arguments provide a background for my approach to the problem of carnivory. However, I do not believe that the feeding of carnivorous animals ceases to be a problem if one adopts different positions. As will be illustrated shortly, feeding meat to animals can create problems even for decidedly non-vegan approaches to human relationships with animals.

Omnivore, Carnivore, Herbivore

Biologists can, albeit imperfectly, classify organisms based upon what they eat. Thus, we can classify living entities variously as carnivores (literally, flesh/meat-eaters), herbivores (plant-eaters), and omnivores (all-eaters). There are a range of subclassifications, such as insectivore (insect-eater) and frugivore (fruit-eater), and some classifications that belong outside of the three most familiar, such as detritivore (detritus-eater). That both scientists and non-specialists are comfortable splitting up the world in this way is philosophically interesting. Implicitly, what living things eat is presented as of primary importance; eating is at the foundation of our taxonomies of the living world.

When we are speaking with precision, these terms – carnivore, herbivore, and so on – do not refer to eating *per se*, but to biological makeup. So, for example, all humans are omnivores. This does not mean that humans necessarily do eat all things, or even that they can eat all things. All it means for a being to be an omnivore is that they can extract some sufficient level of nutrients and energy from *both* (some) plants *and* (some) animals. Until the biological makeup of 'humans' is grossly different from what it is today, humans will remain omnivores. The child who eats dirt does not become partially detrivorous because of that, and the vegan does not stop being an omnivore just because her diet contains no animal products. Consequently, we may worry about nomenclature that contrasts the vegan with the *omnivore*, or, worse, *carnivore*. This language is misleading, not least as it risks naturalizing meat-eating ideologies.

There are obvious differences in practice between vegans and non-vegans, but the difference other than this – assuming both have a choice in their eating practices – is one of ideology. This could be put in terms of the ideology of veganism versus the ideology of *carnism* (Joy 2009).[6] This idea remains disputed, however, and so despite my reservations, I will use *omnivore* to refer to meat-eating humans, noting only that omnivore *qua* biology, omnivore *qua* ideology, and omnivore *qua* practice are three importantly separate categories.

When comparing humans to one another, then, we can meaningfully split them into vegans and omnivores based on their practices, and can meaningfully split them into vegans and omnivores based on their (often hidden) ideological commitments. What we cannot do is split them into herbivores, omnivores, and carnivores based on their biologies. Thus, human-to-human comparisons can be of a different kind to human-to-animal or animal-to-animal comparisons. Compare, for example, a given human with a given rabbit with a given anteater. All three have eating practices, and these practices can usefully differentiate them. Likely, the human (if a Westerner) eats a mix of plants and animals (and more), the rabbit eats plants, and the anteater eats insects.[7] These beings cannot so meaningfully be differentiated by ideology. At most, only the human could be an omnivore or a vegan *qua* ideology. Rabbits, anteaters, and some humans lack ideologies about food (though they may have preferences). The most interesting way to classify these beings *as eaters* is to look to biology, where the three differ quite significantly; the human is omnivorous, the rabbit is herbivorous, and the anteater is carnivorous (specifically, insectivorous). Interspecies comparisons of beings *qua* eaters can, then, take a very different form than human-to-human comparisons – interspecies comparisons can meaningfully differentiate biologically, where human-to-human comparisons cannot.

Given that the *practice* of omnivory/carnivory raises serious moral and political questions, there is a morally significant difference between the meat-consuming animal who is carnivorous *qua* biology and the omnivorous meat-consuming human who is omnivorous *qua* ideology. Philosophers have long been in the business of challenging problematic ideologies. Philosophers are not well situated, however, to challenge problematic biologies, meaning that when problematic biologies are at the basis of problematic practices, philosophers are in a difficult position. Carnivory is thus a problem for philosophers in the way that omnivory, *qua* ideology, is not.

We have thus arrived at the problem of carnivory: carnivores consume (parts of) other animals, seemingly necessitating suffering and death. However, we seem unable to prevent/challenge this consumption, given that it is a consequence of biology, rather than ideology. And this is true whether the carnivore is a being who lives with us in our house and with whom we interact every day, or is a wild animal living on the most inaccessible part of the planet.

Despite my approach to the problem, the ethics of feeding meat to animals is not a question uniquely interesting to those concerned with offering animals greater moral/political standing. There is every chance that defenders of human consumption of animal products will, depending on the kinds of arguments they use to defend meat-eating, have reason to worry about carnivorous animals (Milburn 2015a, 449–50).

For instance, imagine someone made the following argument, which one sees in both the academic literature and popular debate about meat-eating.[8] Animal suffering is a bad thing, and so we cannot inflict suffering without a good reason. A good reason would be a gain in human pleasure that is, relative to the amount of animal suffering caused, sufficiently large. When it comes to meat-eating – continues the hypothetical non-vegan – the suffering of animals in farming is offset by the gains in pleasure that humans experience by eating meat. Therefore, farming animals for meat for human consumption is permissible. But what is striking is that nothing in the argument suggests that *all* beings enjoy consuming animal products to an extent sufficient to outweigh the suffering that leads to their creation. Indeed, given that the argument seems to give relatively low weight to the eaten animal's suffering, it is hard to see why it would give relatively high weight to the animal eater's pleasure – and thus hard to see how it could justify feeding meat to animals.

Or, to use another example, consider an argument drawing on one of the challenges to my argument for veganism alluded to earlier. A certain kind of demi-vegetarian might say that, yes, we need a good reason to kill animals for food, but (much, or some) human meat-eating can be defended on the grounds that it is a form of cultural expression, and expressing one's culture is a good reason to kill animals. But it is not clear that carnivorous animals have this kind of 'cultural expression', or that human cultural expressions depend upon feeding meat to animals. At the very least, arguments to the contrary would be required. So maybe this cultural defence of meat-eating is no good for justifying the feeding of meat to carnivorous animals, even if it can justify some human consumption of meat.

These are merely indicative. Indeed, I think there is work to be done on the assorted defences of human meat-eating, to explore the extent to which they could justify the feeding of meat to animals. It is possible that vegans will find some unlikely allies among defenders of human meat-eating who – if they apply their own arguments consistently – should be opposed to some, or all, practices of feeding meat to animals (or even meat-eating among animals).

Aspects of the problem of carnivory have occasionally been raised by animal ethicists. Some theorists have grappled with (what I have called) the 'predator problem', the question of how to appropriately respond to animals killing and eating each other. Animal ethicists have offered explanations of why humans are not obliged to interfere in these processes (e.g., Everett 2001; Milburn 2015b), or else indicated that animals' predation of each other presents a serious ethical problem that needs, in time, to be solved (e.g., McMahan 2010; Nussbaum 2006, 400-1; see chapter 7 of this book). On the other hand, the feeding of carnivorous companions – though a practical ethical issue confronting any number of real-world vegans and vegetarians – has seen relatively little discussion in animal ethics.[9] These are, perhaps, the most obvious instantiations of the problem of carnivory, but they are – as we will see throughout this book – far from the only ones.

Philosophers are unable to will away carnivory. In addition, and unlike human practices and even pervasive human ideologies, there seems to be no easy way for them to challenge it, undermine it, or call for it to be stopped. If they are to call for the carnivore *practice* of meat-eating to end without something being done about carnivory, philosophers seem to be calling for the end of the carnivore. The problem of carnivory, therefore, is a significant one for animal ethicists, and in a sense trickier than the problem posed by *human* practices of meat-eating. For the remainder of this chapter, I will explore several ways in which the problem of carnivory might be resolved. In offering this examination, I will explore a variety of ways in which carnivorous animals could be fed, some of which are very far removed from how they are typically fed (or how they typically eat) today.

(Not) Feeding Carnivores

As societies and individuals, we feed a wide range of carnivorous animals. Cats, ferrets, and snakes are kept as companions. Tigers, whales, and birds of prey all have sanctuaries devoted to their care. Eager wildlife watchers might choose to feed just about any animal they see visiting their garden. And activists (or well-meaning passers-by) may seek to feed carnivorous animals in the wilderness. Let us assume – for now – that the agents in question are indeed obliged (or at least permitted) to feed these animals. And let us also assume that, if they are feeding them, they are obliged to feed them appropriate food: food which will not make them ill, and diets that will not leave them malnourished. Given that there is a wrong of inflicting death and suffering upon animals to produce food, these agents face a dilemma. Their obligation to the eater appears to conflict with their obligation to the (would-be) eaten. The way forward seems unclear.

Winners and Losers

Two broad options are immediately open to us. Who wins – eater or eaten?

First, we could argue that our obligation to the animals we are feeding outweighs our obligations towards other animals, meaning that we are justified in killing other animals to feed them. This route might be an option in care-based approaches, where personal relationships are lauded – one might have a particularly strong relationship with the animal one is feeding, for instance. However, it has less of a standing in rights discourse. Rights are typically understood as *side-constraints* (Nozick 1974) or *trumps* (Dworkin 1984), meaning that they serve as limits on the actions we may take in pursuit of other goals, even if those other goals are very worthy. Further, negative rights – such as those possessed by the animals who would be made to suffer and be killed so that they can be eaten – are frequently taken to outweigh positive obligations, such as obligations to feed.

Nonetheless, the advocate of killing animals to feed animals might draw upon the resources of the interest-based rights approach to argue that the would-be eater has an interest in being fed, and that this interest grounds a *prima facie* right that outweighs the *prima facie* right of the animals who are being killed to feed them. After all, they will observe, the interest-based rights approach (like many approaches to animal rights) allows cases in which humans *must* kill animals to eat to justify non-veganism. Why not cases in which animals must be fed with meat? The first thing to say is that it remains to be seen whether animals must be killed to feed to other animals. Might there be ways to feed the animals without the need for death and suffering, as there (typically) is with humans? Second, the interest-based rights approach justifies the killing of animals to feed humans in emergencies partly because humans typically have a greater interest in continued life than do animals.[10] When the conflict is between two beings who have a more-or-less equal interest in continued life – say, a dog and a pig who might be killed to feed the dog – there is no reason to think that the (would-be) eater should win out over the (would-be) eaten.

If we aren't going to kill the eaten for the eater, perhaps we could go the other way? It could be that the best way for us to respond to our duty to the carnivores we are obliged to feed is to ensure that no more of them come into existence. That way, we ensure that we are not faced with an indefinitely long future of either wronging animals we kill to feed carnivores and/or wronging the carnivores themselves by failing to feed them (adequately). In an unpublished paper, Katherine Wayne (2013a) addresses the possibility of the 'selective extinction' of domestic cats as a response to the fact that they eat meat. Such an option, I have argued elsewhere (Milburn 2015a), is viable, but only as a last resort.[11] That conclusion may seem alarming; allow me to explain, using Wayne's cats as an example.

Imagine we are faced with two options. One option sees us inflicting a large volume of suffering and a colossal quantity of death on (say) fish indefinitely into the future to create cat food. The other sees us make this generation of cats the last, by ensuring that all domestic cats are neutered. Both cats and fish are rights bearers. Inflicting suffering upon and killing fish clearly violates their rights. Neutering cats (or otherwise controlling their reproductive behaviour to a very high degree) *may* violate their rights – let us assume that it does. What we have is a tragic situation, but, clearly, one in which there is a lesser of two evils. A decade of cat-neutering involves far fewer rights violations (of far lower severity) than indefinite fish torture and killing.

However – both for the cats and for people who enjoy co-living with them – it would be better if there were some third option. Selective extinction is thus a viable but regrettable solution to the problem of carnivory in each case. It is something that we could pursue if there were no other option beyond extinction of the eaters and indefinite animal death to feed them. But we have every reason to hope that there is some other option.

Vegan Carnivores?

Another strategy presents itself. Could we feed animals a diet that does not contain any animal products? This is not a problem when it comes to herbivores. Indeed, it is unlikely to be a problem when it comes to omnivores. Dogs, for instance, can thrive perfectly comfortably on an appropriately planned plant-based diet, just like humans.[12] Of course, the problem of carnivory is an issue precisely *because* we cannot, it is assumed, feed carnivores a plant-based diet.

There is an ambiguity in the word *carnivore*. Among animals classified as carnivores, we can identify those who are *facultative* carnivores and those who are *obligate* carnivores. The facultative carnivore can live as a carnivore, but need not; the *obligate* carnivore is (putatively) unable to live but as a carnivore. As is typical in biological classifications, no clear line separates these groups; facultative carnivores might reasonably be called omnivores, as omnivores exist on a continuum from those who are almost entirely herbivorous (*qua* biology, *qua* practice) to those who are almost entirely carnivorous (*qua* biology, *qua* practice). This fluidity is well illustrated by comparing animals at different times. Evolution is such that traits do not arise or vanish instantaneously, but over many generations. An animal we classify, today, as an obligate carnivore may well be the descendent of an animal – living, let us say, 4 million years ago (mya) – whom we classify unproblematically as an omnivore. Of course, it is not the case that all members of the omnivore species living (say) 3mya suddenly stopped giving birth to omnivore young, and instead gave birth to carnivore young. Instead, if we were to take generations living at, respectively, 4mya, 3.5mya, 3mya, 2.5mya, 2mya, 1.5mya, 1mya, 0.5mya and the present day, we would see animals gradually becoming less like the beings living 4mya and more like those living today. All extant animals exist on a similar scale; the discrete carnivore/omnivore/herbivore classification is merely a shorthand to refer to different segments of this scale.

My point is that we should not be too quick to assume that a given carnivore is of some grossly different kind to a given omnivore, and, consequently, that it is some sort of ontological impossibility that the carnivore could survive without animal protein. It could be that carnivores – even obligate carnivores – can both survive and thrive with a plant-based diet.[13] Domestic cats, again, provide a useful example. Cats' status as a carnivore is stressed repeatedly in classic, widely used works on feline nutrition; the possibility of a plant-based diet is not even acknowledged (e.g., MacDonald, Rogers, and Morris 1984). Nonetheless, and though they are obligate carnivores, some cats are fed diets that do not contain any animal products, typically because of the normative convictions of their human guardians. This provides compelling evidence that (at least some) cats can survive without ingesting animal products. A range of studies of cat diet have come up with varying results about the healthfulness of this kind of diet for cats (e.g., Gray, Sellon, and Freeman 2004; Wakefield, Shofer, and Michel 2006; Dodd et al. 2021), but a 2016 review concluded that a 'growing body of evidence appears to indicate that both dogs and cats can survive and indeed thrive on nutritionally-sound vegetarian and vegan diets' (Knight and Leirsberger 2016, n.p.). The authors noted that 'it is important to remember that dogs, cats – and indeed all species – require specific nutrients, rather than specific ingredients. There is – at least in theory – no reason why diets comprised entirely of plants, minerals, and synthetically-based ingredients (i.e., vegan diets) cannot meet the necessary palatability, bioavailability, and nutritional requirements of cats' (Knight and Leitsberger 2016, n.p.). Other reviews (e.g., Michel 2006) have reached similar conclusions, and even Nestle and Nesheim, whose distaste for companion veganism is clear, allow that cats can live healthy lives on plant-based diets (2010, chaps. 19, 21).

Nonetheless, it would be wrong for me to declare that all carnivores cared for by humans should simply be fed plant-based diets. I say this for three reasons. First, this scientific evidence may well be disputed – it certainly is disputed by many non-scientists, and may be by future scientific work. Indeed, I allow that, as a philosopher, I have no business making authoritative claims about veterinary nutrition. Second, the results may not be generalizable. Cats with certain afflictions, companions of other species, or – most obviously – non-companion carnivores may be unable to survive/thrive on vegan diets. Third, the 'what if' – *what if* there was no prospect of a plant-based diet suitable for a particular carnivore? – makes for an interesting philosophical question.

And, to add to these three reasons, I note that even if (certain) carnivores can thrive on a plant-based diet, we might have reason to want to feed them animal products. For example – more on this in the next chapter – particular companions may prefer meat-based foods.

What other options might be open to us? For the remainder of this chapter, I will consider four: scavenging, eggs, cultivated meat, and animals without rights.

Scavenging

Scavenging for food among humans is associated with the so-called 'freegan' movement, the members of which will take food from bins ('dumpster dive') and perhaps collect animal corpses from roadsides. It is easy to develop a chain of thought that would motivate a vegan to become a freegan. Assume, first, that the wrong of eating animal products derives from the death and suffering of sentient animals, and, second, that our consumption of scavenged animal products in no way supports the death and suffering of animals. It seems, *prima facie*, to follow that consuming these products is not wrong, or, if it is, it is wrong for some other reason than its contribution to the death and suffering of animals. Donald Bruckner (2015) goes further, arguing that the harms involved in arable agriculture (see chapter 5) oblige us to be freegans rather than vegans, on standard arguments for veganism (cf. Bobier 2020).

It is not my purpose here to decide between freeganism and veganism (cf. Abbate 2019a; 2019b; Milburn and Fischer 2021). Instead, I am interested in scavenging as a possible solution to the problem of carnivory. To that end, it is worth noting three related points. First, it is possible that scavenging of food for carnivores does not raise the kinds of health problems that might be raised by scavenging for humans. (Lots of meat that will sicken a human will not sicken a buzzard.) Second, carnivores may be perfectly content with scavenged meat in a way that many humans would not be. (A lion doesn't care that this carcass came from the roadside.) Third – though I concede that the moral significance of this may be slight – it is, in effect, *already* standard for us to feed animals with 'scavenged' material. The animal-derived content of pet food, for example, is often meat deemed unsuitable or undesirable for human consumption, rather than meat that would otherwise be fed to humans (Sandøe, Corr, and Palmer 2016, 221). Thus, feeding carnivores scavenged meat might not be so different from feeding them shop-bought pet food. Indeed, the meat we can scavenge for carnivores could even be better quality – by human standards.

Flesh taken from bins and scavenged from roadsides thus provides a potentially compelling option when it comes to the question of how an individual might permissibly feed carnivores (Deckers 2016, 98–9; Donaldson and Kymlicka 2011, 150–2; Milburn 2017b).[14] At least, this follows if the carnivore we intend to feed is of a species that can be fed using the scavenged food available. Thus, when we are today faced with the problem of feeding carnivores, we might conclude that scavenging represents a viable and morally permissible option. This flesh would either rot away or be eaten by some animal – that animal may as well be our companion, or the resident at our sanctuary, or the visitor to our garden, or what have you.

Accepting this does not commit us, incidentally, to saying that scavenged flesh is morally permissible for human consumption. For example, we *may* – I here take no stance on the issue – have reasons to think that the presentation of flesh as food is morally problematic (see, e.g., Fischer and Ozturk 2016; Turner 2005). However, given the problem of carnivory, we must accept that flesh *is* food for carnivores, even if it is not food for humans. It is telling, for example, that Cheryl Abbate (2019a), in objecting to the idea of humans eating roadkill, argues that the meat should be saved for cats. This is one of the philosophically significant consequences of the problem of carnivory; if we want to eliminate the idea of meat as food, we must surely recognize that we wish to eliminate the idea of it as *human* food. Equally, when we say that dirt is not food, we are saying that it is not *human* food; we are not saying that the detrivore is somehow wrong in eating dirt.

Despite its promise (in many cases) as an individual measure, scavenging flesh is unsatisfying as a lasting political or societal solution. Dumpster-diving is possible only because of the existence of the institution of animal agriculture; our socio-political imperative should be dismantling this institution, not finding new ways to turn the resulting waste into something useful. The collection of roadkill, meanwhile, is viable only for very small numbers (given the relatively low quantity of roadkill), and, in any case, we no doubt have a political imperative to take actions to *limit* roadkill (Donaldson and Kymlicka 2011, 201). If there are many carnivores in need of flesh, roadkill will be unable to feed them. If scavenging is to work as a political solution, it cannot be in the form of dumpster-diving or picking up corpses from roadsides.

We could imagine a system involving the collection of the corpses of animal – nonhuman and, possibly, human[15] – members of our society to produce food to feed carnivorous companions. Imagine it, perhaps, as an extension of existing systems of organ donation. This kind of possibility perhaps sounds deeply dystopic – hardly a respectful way to treat the bodies of our co-community members (Milburn 2020a) – and likely could not provide anywhere near the amount of flesh required. The healthfulness of human and companion flesh for carnivorous animals is also in question. As such, I have little confidence in this kind of possibility. Even while scavenging may provide a stopgap for certain individuals today, the idea of institutionalized scavenging is, I fear, a non-starter.

Eggs

An alternative to carnivores' consumption of flesh could be their consumption of eggs (Donaldson and Kymlicka 2011, 138–9; Deckers 2016, 98–9; Milburn 2017b, 2019a). In short, the idea is that carnivorous animals could be fed the eggs of companion chickens (or other birds); providing the chickens were treated with respect and welcomed – as appropriate – as members of a mixed human/animal society, there is no great harm in their eggs being eaten (Fischer and Milburn 2019; Hooley and Nobis 2016; Wayne 2013b). It is worth acknowledging straight away that this is a proposal touted for the feeding of smaller carnivores. It is one thing to imagine cats or small snakes living off an egg-rich diet. It is quite another to imagine polar bears or orcas living off the same.

The consumption of eggs from so-called backyard chickens is a point of contention in popular vegan dialogue, and sees much discussion in vegan and activist media. Multiple worries are raised. First, there is the fact that backyard chickens need to come from somewhere, and we should not support the industries from which they typically come. This criticism, while indeed viable (it is wrong to financially support hatcheries), does not apply to all cases of chickens being kept by humans. Many vegans keep chickens who have been rescued from animal industry, and it is feasible that chickens kept by humans perfectly respectfully might be ethically bred/choose to breed. (That is, imagine a future society or current sub-society in which chickens are regarded as full and equal members.) This means that chickens could become human companions, and *remain* human companions – just as dogs can remain human companions. And if they remain companions, we could continue to reap benefits from respectful co-relationships with them. Just as we could imagine dogs providing us with the benefits of their company, care, or watchful eyes into the future, so we could imagine hens providing the benefit of eggs.

Second, we might worry about the *use* of these animals. If this is because use is taken to be inherently problematic, then it is the abolitionist challenge. Abolitionists hold that it is illegitimate for us to use animals at all, even if we do so in a way that is consistent with their interests, and even if we do so with every good intention (Francione and Garner 2010). Abolitionism has received much criticism in the political turn in animal ethics; it has been challenged for ignoring the multifaceted and mutually beneficial relationships that humans and animals can have and for problematically assuming that the 'use' of another being is bad for the being (Donaldson and Kymlicka 2011, chaps. 3–4; Wayne 2013b; see also Cochrane 2012). The problem, in short, is that it is difficult to see why using an animal is problematic if the use is in the animal's interests (or, at least, not against them). To take an example from Donaldson and Kymlicka (2011, 134–9), it is hard to imagine what is wrong with a community of humans 'using' sheep to keep grass short; the sheep are happy to live near the humans and to eat the grass, and may benefit from human care in various ways. This relationship could continue generation by generation. Nonetheless, the abolitionist seems to be committed to the claim that the humans are wronging the sheep, and thus owe it to them to end this relationship by preventing another generation of sheep from coming into existence. (This, of course, might involve all kinds of ethically questionable interference with the sheep alive today.) I thus hold that the abolitionist challenge is unconvincing. If the 'use' worry is about the objectification of chickens, then one must ask about the relationship humans have with the chickens. If the chickens are perceived as egg-laying machines, then there is a problem. If the chickens are perceived as companions or family members (or even workers or co-citizens), then there need not be.

Third, there is a worry about the perception of animals (or, in this case, their secretions) as food; again, this challenge may or may not apply when it comes to *human* consumption of eggs, but it should not when it comes to consumption by carnivores. Whether we like it or not, animal protein *is* food for them. Tellingly, one alternative to human consumption of the eggs of rescued/companion chickens frequently touted by vegans is that the chickens themselves consume the eggs.[16] Though chickens are not carnivores, the objection is to *human* consumption of eggs, not consumption *simpliciter*, or to the portrayal of eggs as *human* food, not the portrayal of eggs as food *simpliciter*.

I conclude, therefore, that there is nothing in principle wrong with the eggs of companion chickens being used to feed carnivores – providing, of course, that the carnivores can suitably be fed eggs. We can imagine a household made up of humans, ferrets, and some rescued chickens. The humans would be vegan, and the chickens would be fed primarily with a mix of grains. When the chickens lay eggs, these could be collected, with the animal protein-rich innards being fed to the ferrets and the shells being fed back to the chickens. Everyone can eat healthily and happily. We could even scale this up slightly; imagine a small community in which there was both a chicken sanctuary and a hedgehog rescue. The sanctuary could provide eggs for the rescue, ensuring that the hedgehogs could eat without the need for any animals to be 'sacrificed' for them. Perhaps they could even accept a donation in exchange.

Could this be scaled up further, allowing for an animal rights-respecting egg *industry*? This, I fear, is too large a question for the present enquiry, though it is one that surely requires further thought.

Several animal rights theorists have defended the institution of backyard chickens (e.g., Fischer and Milburn 2019), but, on the whole, they have been more reluctant to defend an egg industry. Some who defend it provide relatively few details about how it could be ensured that the industry respects animals' rights (e.g., Cochrane 2012, 78–9; Zamir 2007, chap. 6); others are open to relatively large-scale egg production (i.e., seemingly more than just a couple of chickens in a back garden), but are reluctant to support commercialization (e.g., Coulter 2020, 35n4; Donaldson and Kymlicka 2011, 138). So, as a political solution to the problem of carnivory, the possibility of large-scale egg production – whether for-profit 'farms', not-for-profit 'community gardens', or something else altogether – remains an open question. The just acquisition of eggs certainly poses a viable *individual* solution to feeding companions (albeit one accessible only to people with the expertise and resources to care for hens), but, even if further work could demonstrate the viability of rights-respecting egg farming, egg production could only ever be a partial solution. While some carnivorous animals might be perfectly happy with eggs, others will not. It is hard to imagine (say) sanctuaries for whales or dolphins successfully feeding their charges an egg-based diet.

Cellular Agriculture

I have identified two viable individual means to feed carnivorous companions – scavenging and genuinely ethical eggs – but I have failed to identify a particularly promising political solution. A viable political solution, I propose, is not immediately available to us. In this sense, today's politics must be about securing tomorrow's solution.[17] The simplest route to this is further research on (particular) carnivores' dietary needs. A greater understanding of these will likely allow us to synthesize plant-based foods that meet their dietary needs. Once again, domestic cats provide a useful case study. Taurine is a nutrient that cats require, and that they would traditionally acquire from meat. However, synthetic taurine can be produced without any animal ingredients, and thus be used to supplement (vegan or otherwise) food for cats – indeed, most commercial cat foods are already supplemented with synthetic (vegan!) taurine (Knight and Leitsberger 2016, n.p.; Nestle and Nesheim 2010, 208).

If carnivores, including cats, are to thrive on a plant-based diet, it is most likely through correct identification of the nutrients they require and these nutrients' production without animals. That said, other factors, such as texture – which may be important for healthy teeth – will also need to be considered.

Research in the right area, however, could provide other solutions to the problem of carnivory. Even if the right plant-based food (or carefully designed systems of egg farming) could be enough for many (say) cats today, the problem of carnivory does not end with cats. Humans find themselves feeding any number of carnivorous animals – carnivorous companion animals; carnivorous animals who venture into human spaces, where well-meaning householders feed them; carnivorous animals living in sanctuaries, rehabilitation centres, and similar, where they rely upon humans for food; and wild carnivorous animals (even those living far from humans) who might be fed by activists and conservationists. And, even if we *were* only talking about cats – famously fussy eaters – a variety of options can only be a good thing.

Perhaps the most exciting prospect for a technological solution to the problem of carnivory comes in the form of cultivated meat. This is meat grown by humans outside of an animal's body. Now, there is lots of discussion in animal ethics and related fields about the prospect of cultivated meat to feed *humans*, or as a stopgap in a transition to a fully vegan world. But what is interesting is that there may be every reason to support the development of cultivated meat even if we are opposed to its use for human food, specifically because it may be valuable for feeding companions (Milburn 2016b, 259n23; Deckers 2016, 99; cf. Ward, Oven, and Bethencourt 2020, esp. chaps. 11–12). Indeed, I believe that the problem of carnivory actually becomes a convincing argument in favour of cultivated meat.

Objections to cultivated meat grounded in animal ethics[18] fall broadly into three categories (Milburn 2016b). First, there are the flesh-as-food objections. This is the idea that cultivated meat problematically presents meat as food (e.g., Cole and Morgan 2013, 211–14; Miller 2012). This is not a worry that we face when it comes to feeding cultivated meat to carnivores; whether we like it or not, meat *is* food for these animals. Second, there are animals-in-the-process objections: the production of cultivated meat still requires some animal input. If the worry refers to the fact that research into cultivated meat has involved the abuse of animals, then this makes cultivated meat no different from many technologies/scientific advances that make our lives better today. Many medical advances, for example, have relied upon vivisection. While we certainly have an obligation not to harm animals any further, we do not have an obligation to roll back to a time in which we did not have these technologies and advances. If we did this, everyone – including animals – would lose out. If the objection is that cultivated meat still requires a minimal number of nonhuman 'donors' in order to be produced, then this seems to be the abolitionist challenge again; we can surely design an institution such that animal 'donors' were able to live happy lives and which did not necessitate the violation of their rights (Weele and Driessen 2013; Dutkiewicz and Abrell 2021). If the objection is about the use of animal products in the production of cultivated meat, then the objection is indeed on to something. Foetal bovine serum, a particularly gruesome slaughter by-product, has been used in the initial development of cultivated meat – but its use is recognized as problematic in the space of 'cellular agriculture' (a term that refers to both cultivated meat and other technologies engaging in 'agriculture' at the cellular level), and the key players are rushing to produce non-animal-based alternatives.

The third kind of objection – and it is worries of this sort that lead Donaldson and Kymlicka (2011, 152) to reject cultivated meat as a viable solution to the problem of feeding carnivorous animals – charges that cultivated meat reinforces a false hierarchy between the edible 'them' and inedible 'us'. The solution when it comes to cultivated meat, I have argued (Milburn 2016b, 256–63), is to not allow cultivated meat to throw up these barriers, and instead permit the creation of cultivated meat made from human cells alongside cultivated meat made from animal cells. Indeed, feeding such flesh to carnivorous animals does not run up against many of the objections that are raised to the consumption of cultivated 'human' flesh by humans.[19] Namely, it is not cannibalism (against which there is a somewhat transcultural taboo), and there is no reason to think that it would be particularly unhealthy. The production of 'human' flesh to feed carnivores even has the advantage of a fully consenting autonomous individual as the ultimate source of the flesh, and so serves to overcome any abolitionist worries we may have about the 'use' of an animal in the production of cultivated meat. It is thus plausible that it would be *preferable* to feed carnivores cultivated human flesh than cultivated (say) bovine flesh – assuming that cultivated human flesh would be nutritious for them. On at least a few occasions, I have found that animal activists' worries about feeding cultivated meat to animals dissolve when it's *human* flesh being talked about (see also Bovenkerk, Meijer, and Nijland 2020).

Cultivated meat is not a viable practical solution for an individual seeking to feed carnivorous animals, simply because it is not yet widely available.[20] When it is available, however, it likely would serve as a highly plausible means for individuals to fulfil any duties to feed carnivores. On the other hand, it does, in a sense, work as a current *political* solution to the question of how we are obliged to feed carnivores, insofar as political communities can (now) fund the development of cultivated meat and (later) take steps to ensure its uptake – at least in nonhuman diets. Cellular agriculture is also (potentially) exceedingly versatile. There is no in-principle reason that we could not reproduce the diet of just about any carnivore, whether it requires the flesh of mammals, fish, or invertebrates, or even non-flesh animal-based foods – blood, eggs, scales, whatever is required. Milk, for example, is already being produced (Milburn 2018), with cultivated dairy products available in the United States. In this sense, technological solutions like cultivated meat provide an extremely versatile and (in the medium term, at least) viable solution to the problem of carnivory.

Animals without Rights

Plant-based diets are viable for many meat-eating animals – including carnivores. But we have now reached a stage where we have viable solutions for feeding carnivores who *cannot* thrive on a plant-based diet. We have solutions for thinking about this as a problem of individual moral obligations to feed particular companions here and now – scavenging and genuinely ethical egg production are viable. *And* we have solutions for thinking about this as a collective political obligation to form a better food system – the development of cellular agriculture offers a brighter future. However, these solutions, it could be pressed, are complex, expensive, and risky. It could be that there is a much simpler solution to the problem of carnivory; namely, it could be that there are some animals who lack rights, and so feeding them to carnivores may be neither morally problematic nor unjust. Such animals could be farmed or in some way collected for carnivore consumption.

Both arguments for veganism that I have endorsed rely upon the notion of sentience. However, the claim that x is sentient cannot necessarily be inferred from the claim that x is an animal. For the most part, this is not an issue; we can be quite certain that many animals who are routinely eaten and fed to carnivorous animals – chickens, fish, cows, pigs, turkeys, sheep, and so on – are sentient. However, edge cases exist. Certain animals utilized by humans as 'sea food' may or may not be sentient; interestingly, and though his position has now changed, Peter Singer (upon giving up meat) originally continued to eat 'oysters, scallops and mussels', believing that 'somewhere between a shrimp and an oyster seems as good a place to draw the line as any' (1995, 174). He later came to adopt a precautionary principle (1995, 174), and I have endorsed something similar (Milburn 2015a). If we are uncertain whether a given being is sentient, better to give her (or 'it') the benefit of the doubt. Other ethicists, touting the potential benefits of entomophagy (insect-eating) – for animals or the environment – have pushed for vegans to recognize that they should be eating insects (Meyers 2013; Fischer 2016a).

These kinds of arguments demonstrate that there has been a shift in the dialogue. Now, the 'question is not whether *any* animals have sentience, but how many' (Cochrane 2012, 24, emphasis in the original). Recent animal ethicists have taken a range of positions here, and some philosophers have argued that insect suffering is potentially a serious ethical problem (e.g., Jenkins 2016; Tomasik 2015), though, as Bob Fischer (2016b) has observed, insect sentience would have radical consequences for animal ethics, potentially necessitating large-scale rethinking. Consequently, others are more conservative.

This is not the place for resolving the many and various ethical problems raised in this area; suffice it to say that these puzzles are difficult, and represent one of the key challenges with which animal ethicists will need to grapple in the coming years. My question, in contrast to many of the issues raised above, is very specific: could we use invertebrates to feed carnivores? Again, part of this will come down to facts about the carnivore. While (certain) fish, reptiles, and birds might be perfectly content being fed (certain) invertebrates, larger mammals may have a more varied response – though it is worth noting that invertebrate-based pet foods are indeed being developed as alternatives to typical vertebrate-based pet foods (Ward, Oven, and Bethencourt 2020, chap. 9).

I take it that the question of whether a particular invertebrate is sentient is primarily a scientific one. Given the existence of scientific disagreement, it is appropriate for us to maintain a degree of conservatism, *both* regarding those animals we label as sentient *and* regarding those we label as *non*-sentient. I am comfortable saying that any animals who (or *that*) are non-sentient are 'fair game'; we can feed them to carnivores. Thus, *if* it is the case that (say) crickets are non-sentient, it seems that there is nothing *per se* wrong with turning them into fodder; and, further, if it is the case that crickets are a suitable source of animal protein for a given carnivore, then we have a neat solution to the problem of how to feed that carnivore. Given my epistemic conservatism, I am not willing to state with certainty that crickets are non-sentient.

I am more willing to say that sponges and jellyfish are non-sentient. This may sound like an empty concession, given that sponges and jellyfish are hardly a useful commodity for feeding the animals the word *carnivore* evokes. But there are animals who eat sponges and jellyfish; if we happened to be feeding one of *those* animals, I would not see any problem with feeding her sponges or jellyfish. In time, as we come to learn more about invertebrate sensation, we *may* be able to say with some certainty that other beings are non-sentient. The non-sentience of crickets, for example, would be highly significant for the feeding of many reptiles. The non-sentience of mealworms would be highly significant for the feeding of many birds.

Given our current uncertainty about sentience in many cases, what we are left with is a large class of plausibly sentient beings (Milburn 2016a). Singer's precautionary principle may be a useful shorthand, but it lacks sophistication. Consequently, animal ethicists have explored a range of techniques for engaging in moral reasoning probabilistically, weighing the harm/wrong of 'hurting'/killing plausibly sentient beings, moderated by the likelihood that there actually is harm/wrong involved in doing these things (i.e., the chance the animal is sentient), against the harm/wrong involved in not hurting them, whether this is in the case of human consumption of these beings, killing them for agricultural purposes, or feeding them to carnivorous animals (e.g., Fischer 2016a; Jenkins 2016; Milburn 2015a). Constraints of space prevent me from exploring the details of these proposals. However, suffice it to say that while we have a good reason not to feed these animals to carnivores *if a suitable alternative can be found*, it does seem that feeding these animals to carnivores will be preferable to either eliminating carnivores whom we are obliged to feed – by whatever means – *or* feeding them beings who are certainly sentient (Milburn 2015a). While the use of plausibly sentient beings to feed carnivores is by no means a perfect solution, it is not a terrible one.

Interestingly, the possibility of feeding non- or plausibly sentient animals to companions may work particularly well at a societal/political level, given that the class of plausibly sentient beings is (at least in theory) temporally limited. Let's say we have a variety of plausibly sentient beings that we use to feed our companions. As we come to learn more about invertebrate sentience, we will come to recognize that these various beings are sentient, in which case we recognize that they are rights-bearers and should not be fed to companions, or else we recognize that they are not sentient, and we have thus found a rights-respecting way to feed companions (Milburn 2015a). Even if we do not feed plausibly sentient beings to companions, we may well have a good reason to endorse this kind of research, as the discovery of definitively non-sentient animals gives us the chance to expand and diversify companion diets – and that can only be a good thing.

Concluding Remarks

The problem of carnivory is a real one, and one that should be of interest to animal ethicists, philosophers of food, and others. Regrettably, it has been widely overlooked. In this chapter, I have diagnosed the problem – carnivory leads to the ethically problematic practice of meat-eating, but cannot easily be challenged or removed, given the biological constitution of carnivorous animals – and explored potential solutions. As individuals who are responsible for feeding carnivorous companions, we can seek out means of feeding that do not necessitate/contribute to the harming of sentient animals. Most obviously, we can ask whether the carnivorous animal we are feeding truly is incapable of surviving on a diet that does not contain animal-derived protein. Many animals are able to do so.

If the carnivorous animal cannot survive on a plant-based diet, we can draw from the existing animal industry without contributing to it, by scavenging waste products. Options unrelated to the animal industry include scavenging from 'roadkill' and utilizing the eggs of companion chickens (or other egg-laying companions). Alternatively, if these methods are not suitable for the carnivore we are feeding, we can seek out non-sentient animals who can be fed to the carnivore, or, as a last resort, we can seek out animals who *may* be sentient, but who we have good reason to believe are not, and feed these to the carnivore.

While these solutions are suitable for lone guardians, they are perhaps not suitable on a wider scale, and may cease to be practical on any scale as food systems change. Political, societal, *collective* solutions to the problem of carnivory are more complex, but more normatively promising and long-lasting because of this. As a society we must, if we are to feed carnivores, take serious steps towards researching alternatives to the current practices of animal feeding. These serious steps, importantly, will remove the need for us to think about the unpalatable alternative – removing carnivory/carnivores altogether. First, and most importantly, they can support the study of animal diets, so that the possibility of carnivores thriving without animal protein can be explored. Second, they can support the development of cellular agriculture. And third, studying animal sentience may allow us to identify animals to which (not *whom*) we owe nothing, and can therefore be used to feed companions.

This research need not, and should not, be limited to biology. Philosophy of mind, for example, has its part to play in the search for sentience, while social science has its part to play in exploring human perceptions of (potential) animal diets. Meanwhile, social and normative thinkers of various stripes are in the best position to diagnose the problem of carnivory and come up with creative solutions to it, as well as best determine the way that we should go about the societal changes that are necessitated by a full understanding of the problem of carnivory.

Research and development in this area, in fact, is already taking place. When I first started thinking about the problem of carnivory in the mid-2010s, I was met with incredulous stares, frustration, and (literal) laughter when I proposed that we should be feeding plant-based diet to carnivores or making pet food out of insects and cultivated meat. However, we can now see a range of start-ups attempting to commercialize invertebrate-based or cultivated meat-based foods for companion animals, in addition to those developing plant-based alternatives (Ward, Oven, and Bethencourt 2020). We even see major pet food producers taking steps to explore alternative proteins; Purina, owned by Nestle, has launched a range called Beyond Nature's Protein, in which pet foods made with plant-based and insect-based proteins are being trialled. (Regrettably, the first products in the range continue to include *some* 'traditional' proteins, but it gives an indication of the possible direction of the industry.)

Do these developments mean that we do not need to worry about the problem of carnivory? Not at all. It means that the practical problems raised by carnivory can be, in the medium term, resolved – assuming that the technology of cultivated meat can develop as it needs to, and assuming appropriate political will to take up this technology as a source of meat for carnivores. Indeed, as noted, the problem of carnivory essentially becomes an argument in cultivated meat's favour. An overlooked and, until now, underdeveloped argument – but not, for all that, an unimportant one. And, as I have suggested, theoretical work remains to be done about how the problem of carnivory interacts with other arguments about meat-eating. It is possible, as I mentioned above, that some ostensibly pro-meat arguments will not actually support the feeding of meat to animals.

But, though they should not make us complacent, these developments give us some reason for optimism. Not only *can* our societies, with good will and thoughtful research, transition to feed carnivorous animals in rights-respecting ways, but crucial first steps have already been taken.

3

Animal Family

Having explored the problem of carnivory, we can now begin to explore the normative significance of the relationships we have with different animals, and, crucially, the positive obligations we have towards them. We begin with the animals closest to us: our animal family, or companion animals. The questions I seek to answer in this chapter can be split roughly into four: (1) Why are we obliged to feed our own companion animals? (2) What moral values should influence our feeding of companion animals? (3) What food-related obligations do we have towards the companion animals of others? (4) In what way should the state be involved with the feeding of companion animals? In answering these questions, I will carve out a distinctive place for companion animals within a wider sociopolitical context; one that recognizes their place in our lives, families, and society, but that nonetheless recognizes the limitations on what should be demanded of us.

A Duty to Feed Our Companions

In order to better understand the food-related duties we have towards companions, we must first ask *why* we have a duty to feed them. I take it that we *do* have a duty to feed our companions, and so – initially, at least – this chapter concerns ethical abduction rather than ethical deduction. I also take it (to point to a different kind of food duty) that we have an obligation to protect our companions from becoming food themselves. By this, I mean that we have a duty to protect companions from predators. While this is important,[1] it will not be my focus.

On one model, our companions are simply our property; if we choose not to feed them, that is our business, no one else's, just as it is our business whether we choose to water our plants, oil our car, or polish our furniture. This model seems questionable – even aside from any commitment to animal rights – insofar as companion animals are disanalogous to plants, cars, and furniture. Companion animals are beings with interests, meaning that things can go well or badly for them in and of themselves. This is unlike plants, cars, and furniture, which (not *who*) can only be said to have bad things happen to them or possess interests in a metaphorical sense.[2]

But the existence of companions' interests is not enough *alone* to show that we have a duty to feed them. If we were to say that it was, we would seem to be committing ourselves to the claim that (in principle) we have a duty to feed many free-living animals with similar capacities (see further Palmer 2010). We could bite this implausible-sounding bullet, and accept that wild animals have strong positive rights to assistance (cf. Cochrane 2013; Cochrane 2018; Sapontzis 1987). Indeed, there are animal ethicists who hold that our general hands-off approach to wild animals might be seriously mistaken (e.g., Horta 2010; 2011; 2013). However, it seems deeply bizarre to suggest that we need to embrace wild animals' equal right to being fed to say that we wrong our companion animals in not feeding them. (For more on feeding wild animals, see chapter 7.)

Though I support determining the (negative) rights of animals by looking to their interests, this, I believe, is insufficient for offering a full and complete delineation of our individual moral duties. Political philosophers are fond of rights-talk, even while moral philosophers need not be. This is because rights concern the use of force, which is the domain of the state. 'Political philosophy', then, 'is not a complete moral theory, nor was it meant to be' (Nozick 1981, 503), as 'some of the ways you ought to treat me, I have no right to, no right to demand or enforce' (ibid., 500). Justice is just one part of the wider domain of normative ethics, but it is a central one; enforcement is closely tied to justice, but not with other areas of morality (see further Garner 2013). I understand an interest-based rights approach to animal ethics to be one key tool for the derivation of political obligations to animals, but this does not mean that a rights approach can demarcate all our ethical (or even political) obligations towards/concerning animals.

To supplement talk of their interests, we must look at the relationship that we have with our companions. Some scholars have talked about companions as objects of love (e.g., Milligan 2010, chap. 6) or as friends (e.g., Townley 2011). Such 'affective' or 'sentimental' relations, we might note (Palmer 2010, 51–4), are particularly useful for grounding human obligations to animals within a framework of care ethics (see Adams and Donovan 1996; 2007), or within a (pseudo-)communitarian ethic (e.g., Callicott 1992; Midgley 1983). However, they are importantly inadequate for making sense of the claim that we are obliged to feed our companion animals.

Why do I say this? Talk of an obligation to feed our companions arises only because some people would (and do) choose not to feed them, presumably indicating that affective/sentimental relationships do not exist. (If someone loves their companion animal, why – outside of a temporary or ongoing crisis – would they not bother feeding them?) Again, we could bite bullets and allow that we are no longer obliged to feed companions if we cease to feel affection for them, but this sounds questionable. To say that we only have an obligation to feed those companions for whom we feel affection is to condemn many animals to being un- or ill-fed, and to remove chance of censure for neglectful guardians.

If affective/sentimental relationships are not suitable for grounding the obligation to feed one's companion, a different kind of relationship is needed. Clearly, companion animals are in a state of *dependency* on humans, and this grounds a relationship between the two. This dependency is not only intense, but enduring. More importantly, however, humans have created this dependency. (Here, I am focussing on the *individual* responsibility of *particular* humans. I will later talk about the *collective* responsibility of humans.) By bringing these animals into our spaces and our control, we have cut off any chance they have of independence (Palmer 2010, 91–2). The situation of companion animals is thus different from those beings who *happen* to be dependent on us – for example, the 'thief' (human or otherwise) who has chosen to live close to us and take food from our insecure store. Instead, we as individuals have chosen to create this dependence. To borrow some helpful jargon from Clare Palmer, in bringing her into our home, we have made it the case that this animal has an *external dependency*: due to the limitations of the animal's liberty (imposed on her from the outside – hence *external*), she has become dependent upon us. This holds whether or not we were involved in the actual conception/birth of the animal. If we were (even if indirectly, because, say, we purchased her from a breeder), then we may well be morally responsible for her *internal dependency* (ibid.). The grotesque breeding of many animals kept as companions means that they are biologically far-removed from their free-living ancestors and/or conspecifics, and thus have little chance of independence (this dependency is thus part of the animal herself – hence *internal*). We will return to internal dependency later in the chapter.

This is not to say, incidentally, that it is necessarily *wrong* of us to foster dependency in companion animals (say, by depriving them of liberty). It is only to say that once we have done so, we have a morally significant relationship with the companion. We can compare the relationship to one between the state and prisoners; one need not commit either way to the ethical propriety of (a particular person's) imprisonment to recognize that once imprisoned, a person has a normatively significant relationship with the state that grounds the state's moral requirement to feed her.

While vulnerability alone may or may not ground a morally significant relationship between individuals, the deliberate creation of vulnerability certainly does. Palmer suggests that the closest human analogy is having a child: 'This choice is widely thought to generate special obligations toward the resulting child, not just because a child is needy and vulnerable, but because *it is your child*; you have obligations toward your own child that you do not have toward children in general' (2010, 94, emphasis in the original). In fact, a closer analogy is adoption (Cooke 2011, 268). In adopting a human child, individuals gain a responsibility to feed that child, deriving from the deliberate creation of a relationship of vulnerability between child and adoptive parents. While the state may be involved in the relationship (insofar as the state oversees adoption processes), and the parents may even have a literal contract with a state-controlled agency, the parents do not primarily wrong the state if they fail to feed the child, they wrong *the child herself*. So, too, in the case of guardians failing to feed their companions, even if they have a contractual relationship with some third party (such as an animal shelter). Thus, just as we have a strong moral obligation to feed a child whom we have adopted, given the creation of a relationship of vulnerability between ourselves and the child, we have a strong moral obligation to feed our companions. Failing to do so wrongs the companion.

What Should We Feed our Companions?

I have established why we have a duty to feed our companions. In the last chapter, however, I explored why we should not feed them most animal products, including much commercially available pet food. It is now time to delve into the factors that may motivate us to feed particular products (including animal products) to our companions. As well as offering much-needed philosophical analysis of the ethical motivations of many consumers – including the motivations that limit the uptake of more ethical feeding practices – this analysis will paint a clearer picture of the nature of the food relationships that we have with our companions. Importantly, such analysis allows us to step beyond talk of obligations to shift away from conventional meat-based pet food.

Food Justice

Food justice is typically taken to refer to questions about the injustice of inter-human relationships in food production and distribution, especially those relating to race and gender. Paradigm examples of food justice issues in production would include the use of child/slave labour in the Global South, the underpayment of undocumented or otherwise vulnerable migrant labourers in the West, and the control – or, through questionable market practices, elimination – of small-scale farmers by multinational corporations worldwide. The reason that food justice concerns are prevalent is precisely that states are failing to protect members of marginalized groups from receiving disproportionate burdens in relation to the food system, and so (in our imperfect world) consumers need to make changes. While states should of course prevent (for example) slave labour, their failure to do so means that individual consumers should act against it – by, for example, refusing to purchase foods gathered using slavery.

The same globalized food system is used to develop, source, create, and distribute food for companions as is used to develop, source, create, and distribute food for humans (Nestle 2008). Thus, companion foodstuffs can face precisely the same kind of food justice issues as human foodstuffs. Given that companion diets do not raise ethical issues that are particularly distinct from human foodstuffs in this regard, it mostly suffices to say that all the food justice concerns that apply to the purchase of human foods *equally* apply to the purchase of foods for companions, whether these are pet foods in the standard sense or not. It is thus appropriate to leave extended discussion of food-justice issues to other writers. (Information about pet foods can be *particularly* difficult to come by, and multinationals can be *particularly* difficult to avoid, when it comes to pet food available on supermarket shelves. But any differences between pet food and human food are, here, matters of degree.)

Concern with the origins of pet foods (among other worries) leads some consumers to favour homemade food for companions. However, to be explicit, there is no reason to believe that only homemade pet food can accord with the requirements of food justice. A study published in the British magazine *Ethical Consumer*, for example, explored the ethical credentials of a range of pet food companies in 2010, recommending several brands – many vegan-friendly (Brown 2010). Meanwhile, a range of ethically motivated pet food start-ups promise a future with a range of ethical options on the table (Oven, Ward, and Bethencourt 2020). Food justice should be a concern – but it need not mean withdrawing from the global food system altogether.

Dignity

There may be an intuitive sense in which bending the food habits of our companions so that they are, for us, convenient, cheap, morally acceptable, or aesthetically pleasing denies the dignity of the companions. It might be thought that such actions present companions as objects to be moulded to suit our visions, rather than as beings worthy of consideration in their own right.[3] Certainly, strands of this critique can be found in popular responses to the idea of plant-based companions. The idea is presumably that certain impositions of food choices on companions are problematic. (The worry cannot be that *any* imposition is problematic; imposition of some kind seems inevitable.)

Thankfully, the putatively undignified nature of certain feeding practices can be mostly put aside (see also Milburn 2017b). First, many theorists are skeptical about the extension of dignity-talk to animals in a way that cannot be attributed simply to speciesism (see, e.g., Cochrane 2010b). But, second, the existing candidates for a philosophically robust account of animal dignity actually offer little support for the suggestion that the way we feed companion animals is problematic.

Some accounts (e.g., Meyer 2001) talk about dignity as something roughly analogous to moral standing – animals matter, and so they have dignity (or else animals have dignity and so they matter). Others (e.g., Anderson 2005) deny that shaping animals to fit into human societies is undignified; indeed, that may be precisely what is necessary to affirm their dignity. Martha Nussbaum (2006), who offers the most well-known and developed philosophical account of animal dignity, explicitly argues that claims about animal dignity should *not* be allowed to override important moral values. For example, her account does not suggest that we are obliged to feed companions meat (Milburn 2017b). Respecting animal dignity, she says, involves protecting the capabilities of animals – allowing them to flourish as the kinds of beings they are. However, 'no constitution protects capabilities *qua* capabilities. There must be prior evaluation, deciding which are good, and, among the good, which are most central, most clearly involved in defining the minimum conditions for a life' with dignity (2006, 166).[4] And though predatory capabilities may be involved in the good life for some animals, that cannot justify harming *others* for the sake of the predatory animal. She praises a zoo for providing a tiger with a ball on a rope rather than an animal to be her prey, concluding that 'wherever predatory animals are living under direct human support and control, these solutions seem to be the most ethically sound' (ibid., 369–71).

Thus, *if* we are concerned about animal dignity,[5] it is not the fact that we are 'shaping' our companions' diets around our values that should be of primary concern. Instead, we should be concerned about the methods of feeding, and the attitudes that motivate them. Our companions should be fed in a way that does not make a mockery of them, and their feeding should not become an object of our derision. Thus, while it is perfectly reasonable for us to enjoy watching our companions eat – just as parents can enjoy watching their children eat, or lovers can enjoy watching their partners eat – there may be something amiss about deliberately feeding them messy foods so that we might laugh at their ineptitude, or feeding them something unpleasant to watch their reaction, or tricking them into thinking there is food when there is not. *These* things can make a mockery of them.

Similarly, while there is nothing in principle wrong with sharing our food with our companions, there could be something problematic about framing them as living bins. Though there is nothing undignified about offering fruit or vegetable scraps to a companion hamster, rabbit, or guinea pig, there could be in coming to view these animals as worthy only of scraps – at least if we take the idea of dignified treatment of animals seriously. Compare: it is one thing to save burnt roast potatoes for a family member who likes his potatoes 'crispy'; it is quite another to assume that one family member need not be treated with the same respect as others and so they can be fed the burnt roast potatoes that no one else wants. The actions (and even the eater's resulting pleasure) might be identical in both cases, but in one case, problematic attitudes about a family member are perpetuated. It is these kinds of attitudes that should be excised – and, even if we are skeptical of dignity-talk, there is surely a lot to be said for this conclusion.[6]

Nature

A second attitude that many have about their companions' diets is that they should be 'natural'. This leads to practices that, from the outside, can seem odd. For example, the vividly named BARF (*Biologically Appropriate Raw Food* or *Bones And Raw Food*) diet, and 'ancestral' diets, seek to replicate the 'natural' diet of companions such as dogs and cats (Nestle and Nesheim 2010, 234–51). Such diets might include raw meat, bones, fur, or whole animal corpses. BARF feeding can be based upon (dubious) beliefs about healthfulness as much as a philosophical commitment to the good of naturalness, but it can be difficult to distinguish the two. Either way, the putative *un*naturalness of companion diets is frequently deployed as a challenge. 'Naturalizing' dialogues are ubiquitous when it comes to the question of feeding meat to companions, even among vegetarians and vegans. 'Narratives of human choice', Erika Cudworth explains in a qualitative study, 'infuse practices of [human] dietary resistance to carnism, whereas biologism effectively reproduces the "love/eat" distinction when it comes to feeding animal companions. As one interviewee put it, "I'm a vegetarian, but she's a dog"' (2016, 223). This is ultimately attributable, says Cudworth, to carnist modes of thought, with guardians 'drawing on arguments for biological difference and psychological necessity' (ibid., 239). And, perhaps unsurprisingly, there is near-ubiquitous talk about 'naturalness' in pet food branding and advertising (Nestle and Nesheim 2010, 130–2). As such, I take it that it is worth critically examining the suggestion that companions be fed a 'natural' diet, even if such an idea may not be particularly popular in philosophical dialogue.

There is something paradoxical about the suggestion that companions – especially dogs and cats – be fed a natural diet. This is because the companions themselves are so far removed from 'nature' that the whole idea becomes meaningless. Many environmental ethicists have stressed just how unnatural companions are. John Rodman, Holmes Rolston III, and (in his earlier work) J. Baird Callicott argue that companions are problematic precisely because *they* are unnatural, or have been denaturalized, and so have become 'living artifacts' (cited in Cochrane 2014, 158). There is something curiously selective in moralizing about companions being fed 'unnatural' diets while not also criticizing practically every aspect of companionship, right down to the physical makeup of companions themselves:

> [Guardians] frequently microchip, vaccinate, de-worm, de-flea and de-sex their animal companions, and confine them indoors at night because they correctly believe such steps are recommended to safeguard health. Clearly, such owners are willing to depart quite radically from 'naturalness' when they

believe it may protect the health and welfare of their pets. Accordingly, the resistance of such owners to the concept of vegetarian companion animal diets is more likely to stem from ignorance about the hazardous ingredients found within commercial meat-based diets, and about the potential of nutritionally-sound vegetarian diets to safeguard health, than from any deep-seated commitment to 'naturalness'. (Knight and Leitsberger 2016, n.p.)

Some companies producing pet food will lean upon misleading or even deceptive claims and imagery about the 'naturalness' of their product, or the 'naturalness' of a meat-containing diet for companions. In so doing, they 'completely erase the reality of how [companions'] nutritional needs are met today' (Wrye 2015, 109), playing on a tension between the idea of companions as ferocious killers, but also as vulnerable, loved family members. Interestingly, the meat content of commercially available pet foods is one thing that the manufacturers are keen to hide, given that these products may have a higher plant than animal content (ibid.).

Individuals putatively concerned with the naturalness of particular animal diets also face the extremely difficult task of identifying, first, what constitutes the 'natural', and, second, what is so good about it (Milburn 2017b). These puzzles are already notoriously difficult when it comes to human food (see, e.g., Dussault and Desaulniers 2019; Siipi 2008; 2013). Put simply, it is unclear what level of human intervention in food production makes something unnatural. Agriculture, for example, is surely not a 'natural' phenomenon.[7] Presumably, though, advocates of 'natural' eating do not oppose the consumption of all products of agriculture, so an account is owed of what makes agriculture itself sufficiently natural, even while some new innovation in agriculture is not.

It is also unclear why naturalness is something to be praised. There are lots of things apparently natural that we think of as bad things (disease and suffering, perhaps), and other things apparently unnatural that we think of as good things (hospitals and libraries, perhaps). These difficulties extend to pet food (Nestle and Nesheim 2010, 130–3), and are only compounded by the introduction of questions about the naturalness of companions themselves. I conclude, then, that we do not need to be worried about the naturalness of the food that we feed to our companions, or the naturalness of their eating practices. The concept is not a useful one when it comes to the ethics of our food-related practices involving our companions.

My rejection of the virtue of 'naturalness', however, certainly does not entail that we have no *environmental* obligations concerning our companions' diets. Indeed, I have elsewhere argued (2017d) that any concern with 'naturalness', given the overwhelming 'unnaturalness' of animal agriculture and the fact that it is environmentally catastrophic, should lead us to feed companions plant-based diets *even if* such diets are themselves 'unnatural'. The environmental impact of companion animals should lead to us rethinking our feeding practices as a part of a wider obligation to reduce the environmental footprint (pawprint, hoofprint) of ourselves/our families/our societies/our states (Milburn 2019a; Ward, Oven, and Bethencourt 2020, chap. 2). A rejection of the moral significance of 'naturalness' (in this case) does not equate to a rejection of the pressing normative obligations we have concerning the environment.[8] As with food justice concerns, I do not see that companion animal diets raise any distinctive concerns when it comes to environmental impact, so exclude an in-depth discussion here.[9]

Freedom

The possibility that companion animals be offered a degree of freedom to autonomously choose their own diets is one that is both practically and theoretically contentious. On a theoretical level, debate rages about the extent to which animals might have an interest in autonomy, and in what way (see, e.g., Blattner, Donaldson, and Wilcox 2020; Cochrane 2009; Giroux 2016; Healey and Pepper 2021). On a more practical level, animal ethicists – even those endorsing plant-based diets for companions – exalt the idea of offering companions choice over their food:

> It's true that humans need to ensure that dogs meet their nutritional needs, and that they don't overeat, or eat foods that will poison them. But this still leaves a large area in which dogs can express their food preferences and make their own

choices. Through trial and error (and choice amongst options), it became perfectly clear to us that our dog Codie's favourite foods included fennel, kale stems, and carrots. And peas were so prized he simply helped himself from the veggie garden. Fruit really wasn't of interest. On the other hand, his buddy Rolly was mad for bananas. Dogs have individual preferences, and (to varying degrees) the competence to make choices based on their preferences. (Donaldson and Kymlicka 2011, 109)

Empirical studies on the attitudes of guardians illustrate an ambivalent relationship to companion choice when it comes to meat-eating; on the one hand, companions are portrayed as unlike humans in that they lack a choice over whether to eat meat (Cudworth 2016, 235), while, on the other, companions are portrayed as 'choosing' to eat meat, so that feelings of guilt and disgust relating to companions' food can be overcome (ibid., 237). In both cases, however, talk of companions' (lack of) choice serves to obfuscate the choice of the *guardian* to feed meat to companions.

Similar obfuscation seems to be present in the work of the animal ethicist Jessica Pierce, who argues for the importance of giving companions choice in their diet by seemingly suggesting that guardians have an obligation to feed companions *preferred* foods. Pierce makes this argument to defend her choice to feed her companions meat, even though she indicates that she is a vegan, and notes that she raised her daughter vegetarian (2016, 79). She writes that, in addition to having worries about the healthfulness of plant-based diets for companions, she 'must admit to another factor in my decision making: the pleasure my animals take in their meals. For a period of several months, I did make my dogs go vegetarian, and briefly vegan, and I suffered many looks of disappointment and approbation when their bowls were presented to them at meal times.' On the other hand, Pierce draws 'a line' at those companions who are typically fed live food, or even dead-but-whole rodents. It is not obvious that there is anything more morally odious about these practices than feeding companions chunks of meat of uncertain provenance. Consequently, and by her own admission, her willingness to feed companions meat rests upon her ability to distance herself from the killing of animals (2016, 80).

Given the arguments of the previous chapter, it should not be surprising that I am unconvinced by Pierce's claims. We should not accept arguments about how humans enjoy eating animal products in defence of human practices of meat-eating, so I do not see why we should accept claims about dogs' enjoyment of meat in defence of feeding meat to these dogs. Compare: we would not be particularly impressed with the argument that (particular) dogs enjoy chasing rabbits more than chasing sticks, and so guardians of (particular) dogs may purchase rabbits for their dogs to chase instead. Pierce's concern, I suspect, rests more on her own failure to successfully transition her dogs to a plant-based diet. There are practical steps that can be taken. Gradual changes, mixing new food with old, can help ease a transition (Gillen 2003, 93–101). Further,

> [guardians] should clearly demonstrate that they consider the new diet [to be] just as edible as the old (without possibly warning or alarming their [companion] by making a fuss). They should not be concerned if animals eat around new food at first. Simply having it in close proximity will help create the necessary mental association, as will mixing the food thoroughly. The addition of odiferous (the sense of smell is very important) and tasty additives, such as nutritional yeast, vegetable oil, nori flakes and spirulina, can all help, as well as gently warming the food. Offered food should always be fresh. Gradual change and persistence are the most important factors for transitioning resistant animals, however. (Knight and Leitsberger 2016, n.p.)

I take it, then, that companions' own choice/preferences cannot be decisive when it comes to feeding practices in conflict with important other obligations possessed by guardians, such as obligations to third parties – especially animals who would be killed to produce food. They should certainly be considered, insofar as animals' own choices about food indicate which foods they prefer. All else equal – and though we must not jump to too-quick conclusions about what the animals *do* prefer (Ward, Oven, and Bethencourt 2020, 141–7) – it is better to offer preferred food than non-preferred food. But when all else is not equal, companions' preferences are just one consideration among many.

Paternalism

Let us return to the idea mentioned earlier, that 'humans need to ensure that dogs meet their nutritional needs, and that they don't overeat, or eat foods that will poison them' (Donaldson and Kymlicka 2011, 109). This points to a broad class of reasons for which we may be justified limiting our companions' choices concerning what they eat and/or overruling their preferences beyond moral duties to third parties: such interference could be in the interests of the companions. Such paternalistic interferences in the lives of our companions do not seem to be merely permissible; they seem to be required.

I began this chapter by addressing the idea that it is wrong not to feed our companions. However, this, in a sense, emphasizes the wrong challenge. When it comes to feeding companions, obesity is a far greater problem than underfeeding (Nestle and Nesheim 2010, chap. 15; Pierce 2016, chap. 19; Sandøe, Corr, and Palmer 2016, chap. 8). As in humans, obesity in companions leads to serious health and other welfare problems. Companion obesity is – at least partly – down to us.[10] We decide whether our companions have access to food, and both the food's content and its volume. We confine our companions, limiting their opportunity for exercise, and potentially introduce psychological problems that lead to stress-eating. And we feed treats to our companions. Like human junk food, these items have high fat, salt, and sugar content and little nutritional value.

We are thus generally responsible for making our companions obese. Given our companions' already established relationship of vulnerability to us, and given the negative welfare effects of companion obesity, we have good reason to believe that in making our companions obese, we wrong them. We are not so much faced with a choice about *whether* to control companions' diets, but *how* to do so. This again speaks to the analogy between our companion animals and our dependent family members, such as our (adopted) children. We have a responsibility to ensure that our children's diets are healthful and appropriate, even if the children themselves would freely choose a diet of junk.

On the other hand, there are some individuals whose preferences we would not be permitted to override in the name of their own good. We could not override the autonomy of adult members of our family, roommates, tenants, or live-in employees – or not nearly in the same way that we might override the autonomy of our children/companions. I cannot imprison my brother to feed him only healthy foods, even if I am convinced (even if I am *correct*) that doing so would be for his good. However, this is (close to) what parents and guardians might reasonably be permitted, even *required*, to do concerning dependent children and companions.

The difference here stems from a combination of factors. The first is the relationship of vulnerability; this is what makes the guardian (as opposed to anyone else) responsible for the companions' diet. The second is the fact that the companion herself may have only a limited ability to regulate her diet in accordance with her interests. For example, the companion's 'wild' ancestors may have evolved to gorge themselves with food at every opportunity; in an environment of scarcity, this is both viable and sensible, but in an environment where food is highly plentiful, this strategy will see the companion quickly ill.[11] Paternalism is viable when it comes to companions, then, for the same reasons that it is appropriate when it comes to children and adults incapacitated in a variety of ways.

This is not to deny that there may be ways in which animals possess the ability to engage in autonomous choice when children and incapacitated adults do not. My argument is not that (companion) animals may always be treated paternalistically, only that guardians are obliged to protect companions from their own destructive eating habits by feeding them an appropriate, healthful diet, even if that is not what the companions themselves would choose.

Guardians, then, neither (necessarily) have an obligation to respect companion choice/autonomy/preference in cases when doing so conflicts with duties to third parties, nor when doing so would go against the interests of the companion herself. This still leaves a large space for respecting the autonomous food choices of companions. However, and in contrast with the kinds of claims implicitly suggested by Pierce on the one hand and Donaldson and Kymlicka on the other, I am unconvinced that we have sufficient reason to believe that guardians are *required* to respect companion choices/preferences in this way. This would entail that guardians who choose to feed a perfectly appropriate and healthful diet to a companion who, as it happens, would prefer a slightly different appropriate and healthful diet are doing something wrong, and this is not a claim that we should make. It would undoubtedly be good of a guardian, *ceteris paribus*, to facilitate a companion's dietary choice insofar as she is able, but such a thing would be supererogatory – though good, it is required neither by justice nor morality. So, for example, and though I do not commend it as ideal, I see no necessary wrongdoing on the part of the guardian who feeds their dog ample, healthful food for the whole of their relationship, but never offers her a treat, even though it is known that the dog likes treats. This practice does not necessarily warrant censure, unlike the case of the person who, perhaps because of her love for her dog, feeds *too many* treats.

Once again, the analogy between companions and young children is apt. Parents who fail to give their young children a choice over their food may well be viewed as strict, and those who are not concerned (beyond a basic level of palatability) with children's preferences about food might be considered puritan. However, such parents should not be the target of disapprobation. Condemnation is rightfully reserved for those parents who fail to feed their children, or else feed them extremely poor-quality foods, resulting in diets that leave the children malnourished, or obese, or sickly.[12] So too with companions; they are wronged if fed badly, but not if fed well with things they would not have, themselves, chosen.[13]

Other People's Companions

I have identified why we have an obligation to feed our own companions and sketched, in broad terms, factors that will and will not go into determining the ethically appropriate feeding. Before turning to questions about the role of the state in companion animals' diets, it is worth briefly considering the kind of food-related moral duties that we may have concerning other people's companions.

First, we will normally have obligations concerning the companions of others if we have a particular kind of relationship with those others. There is a degree of transitivity in relationships of obligation created by vulnerability. For example, imagine Smith lives alone with her ten-year-old daughter and her daughter's companion lizard. It seems fair to say that Smith has a responsibility for that lizard's diet. Her daughter is vulnerable to her, and her daughter's lizard is vulnerable to her daughter, meaning that the lizard is vulnerable to her. If – as may be appropriate, depending on circumstances – Smith forbids her daughter from leaving the house or takes her away on a trip, Smith's daughter may be unable to acquire food for and/or feed the lizard. It is true that Smith plausibly did not choose to enter into this relationship with the lizard in the way that she (presumably) did with her daughter, but her daughter's acquisition of the lizard is, in a meaningful sense, her responsibility; she is obliged to look out for her daughter and keep tabs on her activities (to a degree appropriate to her age and level of responsibility), and consequently Smith has (even if only passively) 'permitted' her daughter to acquire this lizard.

In this scenario, Smith's daughter's continued possession of the lizard is, in a large way, down to Smith. If such a thing was not too harmful for the lizard himself, Smith could terminate her relationship with him by finding another responsible adult to take care of him. (Normally, though not necessarily, this would also sever the daughter's relationship with the lizard; both her interests, and the lizard's interests, should be considered in this decision.) However, until steps like this have been taken, Smith has a degree of responsibility for her daughter's companion's diet, even if her relationship with her daughter's companion is not *as* voluntary as the relationship she could have with a companion of her own.

The above duties to the companions of others derive from the *moral* duties that others have concerning their companions. There will also be derivations from the duties of justice owed to companions by the state. I will discuss these in depth shortly. For now, we can note that if the state has a duty to protect companions' rights to be fed, we (as individuals) will likely derive certain duties from this, even when the companions concerned are not our own. Most minimally, we will have a duty not to get in the way of the state when it acts to protect these rights – we should not protect our neighbour's 'right' not to feed his companion. Less minimally, we may have an obligation to report wrongdoing. If we are aware that our neighbour is not feeding his companion, then we should notify the relevant authorities. Most controversially, we may have an obligation to act if the state fails to do so. If we live in a society in which companions' rights to be fed are not (adequately) protected by the state and our neighbour is failing to feed his companion, we may have an obligation to seek some way to protect the companion. Feeding her ourselves, adopting/purchasing her from our neighbour, or even – at the extreme – 'rescuing' her may be appropriate. Interfering in the lives of others in this way, however, raises difficult issues, and can easily be done wrong. The sociologist Lesley Irvine (2013), for example, opens her book on the companions of homeless people with an anecdote about a time she and an animal-loving friend – with every good intention – offered to purchase a dog from a homeless man after their attempts at having the authorities take the dog from him had failed. Their efforts, she illustrates, stemmed from mistaken beliefs about the inability of homeless people to care for their companions.

Talk of homelessness and animals, however, raises another possibility; it may be that we have obligations to help others to carry out their duties towards their companions. This is a different kind of case from the one above, as it concerns a situation in which a guardian *wants* to feed their companion, but is, for whatever reason, unable to do so. Steve Cooke (2011) expands on a particular kind of relationship that exists between us and the companions of others; the companions are the property of these others, and, consequently, we may 'have Good Samaritan duties to protect the companion animal owner's property' (2011, 266). Cooke's arguments can apply meaningfully even if we wish to reject talk of companions as property. Given the significance of companions in the lives of many of the people to whom we are (in some morally relevant sense) close, we have indirect moral duties to loved companions to feed them if they are in need and to prevent them from becoming food. (Again, the analogy with children is apt.)

This may sound insignificant, but it is not; it establishes a relationship between me and many companions with whom I have no affective relationship and who are not particularly vulnerable to me. In practical terms, it gives me a clear reason to believe that I may have some reason (even if that reason does not, alone, necessarily ground a moral obligation) to offer food for the dog of a poverty-stricken person[14] or shoo away a fox I see eyeing a neighbour's rabbit hutch, *even if* I am unconvinced that I have a comparable reason to feed a hungry dingo or shoo a fox away from wild rabbits. This kind of obligation will be moderated by the kind of relationship I have with the other humans in question; thus, I may well be morally *obligated* to feed the companions of my close friends and family (to the extent that it would be wrong of me not to) if, for whatever reason, my friends and family are unable to do so and the companion is in need. On this picture, it would be good of me to help acquaintances and strangers by feeding their companions if they are in need, but – absent some other obligation – it would be supererogatory; I should not face censure for failing to do so.

As this last line indicates, I do not think that this is the last word on the duties we have towards/concerning the companions of others. To start to explore this, however, we need to move from questions of moral entanglement to questions of political entanglement.

Companion Politics

My focus in this chapter so far has been on companion diets/feeding as a moral problem. However, a focus upon this alone would leave my account impoverished; normative questions about companions' food and diet are not solely questions about individuals' relationships with (particular) companions, but about the relationship of companions collectively to the state and about the obligations that the state has towards/concerning companions. The pertinent questions can be roughly split into three: (1) Human-focussed regulation; (2) Rights-derived protection; and (3) Membership-based distribution. I will address these in turn.

Human-focussed Regulation

Human-focussed regulation refers to those legislative moves that are made to protect consumers from questionable and problematic elements of the food system. Existing and possible regulation of this sort is a recurring focus of the food studies scholar Marion Nestle, author of *Pet Food Politics* (2008) and – with Malden C. Nesheim – *Feed Your Pet Right* (2010). In these studies, it is revealed that one cannot separate the animal feed (including pet food) industry from the human food system; the structures are one and the same. This means that the human food system cannot be adequately regulated (and the human food supply cannot be adequately safeguarded) without similar legislation in the pet food system. For Nestle (2008), companions can be Chihuahuas in the coal mine; like the idiomatic canaries, their deaths can be alarm calls about impending crises in the human food chain. In *Pet Food Politics*, Nestle shows in detail how the 2007 revelation of melamine-contaminated pet food was a precursor to the realization that melamine had entered the *human* food chain; contaminated foodstuffs ended up being fed to pigs, cows, sheep, fish, chickens, goats, and shrimps who were intended for human consumption (Nestle 2008, chap. 15). More broadly, the melamine crisis revealed the extent of the problems faced by the globalized food structures on which contemporary Western food systems – including the systems providing foods for humans, companions, and farmed animals – stand.

Despite this, the pet food industry is alarmingly under-regulated, with little legislation and even less enforcement protecting consumers from misleading or even fraudulent food- and supplement-labelling claims. Nesheim and Malden present a series of (US-centric) recommendations (2010, 307–13). They make requests of the pet food industry, the rendering industry, the American Association of Feed Control Officials, the United States Food and Drug Administration, the United States Department of Agriculture, veterinary schools, and guardians. Many of those directed at the legislative/pseudo-legislative bodies focus strongly on information provision, while many of those directed at guardians involve *requesting* information provision, and many of those directed at veterinary schools involve *teaching* about the industry. The authors seemingly suggest that consumers (in this case, the consumers are the guardians, not literal consumers of food) have a *right* to information, and a *right* to protection via appropriate legislation.

Thus, Nestle's (and Nesheim's) calls for greater regulation of the pet food industry is based ultimately upon inter-human justice concerns, and the interests of humans. First, the pet food industry should be regulated to protect human food supply, and, second, the pet food industry should be regulated to protect the interests of those who are purchasing its products – guardians. My point here is not to criticize Nestle and Nesheim. Guardian-focussed regulation, as they hold, can indeed help companions as well as guardians; a guardian protected from fraudulent health claims on pet foods is more likely to seek out genuine veterinary care for a companion who needs it, for example. In addition, Nestle and Nesheim are right that there are serious normative concerns – even aside from duties to animals – about the way the pet food industry operates. Consequently, I second their calls for tighter regulation of the pet food industry to protect the (human) public from fraudulent/misleading claims, and protect animals (including humans) from questionable food-production practices, understood broadly.

Rights-derived Protection

Human-focussed regulation will be only one small part of the wider policy considerations that should feed into law making when it comes to the feeding of companions. Of greater interest and significance are the rights of companions themselves; they, to a greater degree than their guardians, are affected by and reliant upon laws and other regulations concerning their feeding. I have argued that the interest-based rights approach is well-suited for demarcating the negative rights of animals, and – because sentient animals have an interest in not being made to suffer and not be killed – it is easy to conceive of rights generally possessed by companion animals against being poisoned, or otherwise fed foods that would make them sick. However, this is not a conclusion unique to companions; *all* sentient animals would generally possess rights against being fed foods that will kill them or make them suffer.

What is more normatively interesting is the possibility that companion animals possess a *right* to be fed. This is distinct from the already-settled question about guardians having a moral duty to feed their companions, as it would entail that a companion who is not fed is the victim of an *injustice*. If – to echo earlier comments – a failure to feed companions was an injustice, and not merely immoral, then those failing to feed their companions could (should!) face legal censure. The trouble is that justifying such a right is potentially tricky.

If we are concerned with negative rights, it may seem that we could posit a right against neglect. This is problematic; it opens the door to free-living animals possessing a right to be fed. An advocate for wild animals could observe that we surely do 'neglect' wild animals by failing to take their suffering seriously, as we do – putatively – when we dismiss the claim that we have an obligation to feed them. (Perhaps wild animals do have a right to be fed. But it is surely a mistake to say that companion animals can only have a right to be fed if wild animals do.) Another possibility would be a kind of conditional right. Animals possess a right to be fed *if* they are not sufficiently left alone. We could thus say that free-living animals have no right to be fed for as long as they are left alone, while companion animals do have a right to be fed, as they are not being left alone. This approach starts to sound promising, at least insofar as it gives us an intuitively plausible result. Indeed, I defend something similar, and the possibility will be returned to below. However, it has its problems. Not only does it sound suspiciously *ad hoc*, but it might seem to entail that we could absolve ourselves of a duty to feed companion animals by simply abandoning them.

Let us take a step back. Consider why we might think that companion animals possess (something like) a right to be fed. First, there are the already-discussed relationships of dependency. These mean that companion animals have no way to feed themselves; they are confined in human homes and spaces. (Of course, their confinement need not be understood pejoratively [see Donaldson and Kymlicka 2016].)

Second, there is the fact that companion lives have been shaped by human contact. In paradigm cases, companions *themselves* have (literally) been shaped for human purposes: dependency is, as was noted above, internalized (Palmer 2010, 92). What this means is that dependency has been bred into the animals in question. There is no comfortable home for a pug, for example, separate from a mixed human/animal society – even while there is for her conspecific wolf cousins. Even in cases of relatively 'wild'[15] animals kept as companions, dependency – itself a relative concept – can develop, and *become* internalized. This means that there is, to a greater or lesser extent, no way that a companion can be 'rewilded'. There is no human-free environment in which companion dogs, companion cats, or even companion horses can live and thrive in the way that their wild ancestors could, or wild cousins can.

Third, there is the fact that moral agents (i.e., particular humans) are responsible – both causally and morally – for the plight of companion animals. This last point is significant; entities/beings lacking in moral agency, including natural phenomena and all (actual) nonhuman animals, cannot violate rights. I have no rights-based claim against the bolt of lightning that strikes me; if I wish to appeal to rights, I will have to claim that I have a right to assistance or protection. The closest I could come to claiming that my rights have been violated by the bolt of lightning is to find some agent who is morally responsible for the lightning's creation – science-fiction scenarios aside, this would be difficult – or morally responsible for my presence in the thunderstorm.

Where does this leave us? Companions as individuals are vulnerable to humans as individuals (or small groups). What is more, this vulnerability is *irreversible* – it is here to stay, at least in the short term. There are no precise human analogues, but small children are perhaps the closest. With small children, it is relatively uncontroversial to say that the state is obliged to ensure that their guardians provide a baseline level of care, including, but not limited to, feeding them a minimally appropriate diet. If the parents fail to do so, it is incumbent upon the state to act. I propose that this extends *mutatis mutandis* to the guardians of companion animals.[16] The kind of rights in the offing here are the basic rights that sentient animals possess against the infliction of suffering and death, and, given the already-established obligations possessed by guardians to feed companions, it is meaningful to talk about the suffering and death resulting from a lack of feed being *inflicted* upon companions by guardians.

There is another sense in which companions are vulnerable to humans, and much will be made of this in the next section. Specifically, companions *as a whole* are vulnerable to humans *as a whole*. Companions as a whole have been incorporated into human communities, and have then bred alongside – typically, directed by – humans to have certain features associated with companionship (or work). Thus, companions *generally* have both external and internal vulnerabilities to human communities (or, if preferred, *the* human community). Humans, collectively, have the lion's share of responsibility for this. Just as bringing an individual animal into our home cuts off a range of possible futures for that animal and makes that animal vulnerable to us, bringing a whole population into our society and cutting off alternative possible futures for all of its individual members makes each of these members vulnerable to us *collectively*. And this broader vulnerability grounds obligations in much the same way that the narrower vulnerability does – *all* companion animals, even those without an obvious guardian, thus have claims to be fed because of their relationship of vulnerability to human society more broadly. (This is in principle compatible with the collective identifying *particular* humans who have a responsibility to feed those without any obvious guardian – say, breeders – or taking steps to ensure that there are no such animals, such as through sterilization programmes.) The vulnerability of companion animals is not, then, a simple matter of the relationships that particular animals have with particular humans – it is much more fundamental, and much broader, affecting far more animals, including humans. Companions are now almost inextricably 'tainted' with human agency; while they continue to exist, we owe them some basic level of care.

The plight of starving companions (whether beloved family members, homeless, or living in a shelter of some kind) is thus relevantly different from the plight of starving free-living animals in paradigm cases. Free-living animals are not vulnerable to humans in the way that an (adopted) infant is vulnerable to her parents, but in the way that an adult is vulnerable to potentially violent strangers who pass her on the street: if these strangers choose not to interact with her, her vulnerability has not been translated into harm. If parents choose not to interact (even indirectly) with their infant children, however, the infant has been harmed. By choosing not to interact with free-living animals in a significant way, we leave them free to feed themselves to the best of their ability.

Wild animals left alone are obviously relevantly unlike dependent family members (human or companion) or animals living in a shelter. But they are *also* relevantly unlike homeless companions, or 'strays'; this is because the homeless companion continues to possess a level of internal dependency, while the paradigmatic free-living animal who is starving does not. Indeed, even if the free-living animal is 'dependent' upon and 'vulnerable' to humans insofar as she will die if not fed by humans, she differs significantly from the companion – even homeless companion – in that this dependency (in paradigm cases) is not caused by humans. (Even the companion who has never lived with humans – the puppy of strays, for example – has *internal* dependencies that the free-living wolf puppy does not.) This means that no human (or human collective) could be held responsible for the wild animal's suffering and death, and thus that no negative rights violation is taking place. (The significance of harm being attributable to humans, recall, is that many humans are moral agents. Species membership is, strictly speaking, irrelevant.)

We thus see that something like a conditional right possessed by animals to be fed if not sufficiently left alone is not as *ad hoc* as it may have seemed, and is even able to ground the rights of many homeless companions once we properly understand what is meant by 'left alone'. Thus, we can meaningfully talk about companions' 'right to be fed' in a loose sense even while sticking to the idea that the interest-based rights approach is suitable only for the demarcation of negative rights – once we begin to interfere with animals, imbuing them with dependency, we have violated their negative rights if we do not offer them the resources and protection they need during the period of interference. And once we (collectively) have interfered with animals *so much* that their dependency has been internalized, there can be no end to the period of interference.

What does all of this mean in practice? Maybe we could envision a system of licensing, according to which people may only take on a companion if they are approved by the state (Cochrane 2012, 129–36). Part of this approval process would be a demonstration that the prospective guardians have the resources, ability, and expertise necessary to adequately feed said companion. It would also involve, presumably, the legal acceptance of responsibility on behalf of the guardian. The guardian would allow that they could not simply abandon the companion, or fail to feed her, or fail to take reasonable steps to protect her from being eaten herself. Maybe said guardian could be required to take out insurance for their companion; this insurance could fund the feeding of the companion if the guardian finds that they are no longer able to do so themselves, or if negligence leads to the companion being taken out of the guardian's care. This system would give the state a clear and unambiguous mandate to seize mistreated companions; this could be the action of last resort in cases when they are not being adequately fed, just as it presently is for children. In the event of companions being seized from neglectful guardians, the state would enter into the same kind of relationship of vulnerability with the companion as exists in a guardian/companion relationship; thus, the state would take on an obligation to feed the companions. (This could be funded by the license/insurance system, thus sidestepping objections from those who resent their tax money being used to feed companions.)

A system like this has the advantage of being readily implementable. Indeed, similar systems are already in place in some jurisdictions, and there are comparable systems in use among many private institutions offering companion animals for adoption. But it does not capture everything that is significant about the rights of companion animals to food. For example, it does not offer a full account of what is owed to guardianless companions – they are vulnerable to humans collectively, but what this means remains unclear – and nor does it offer any means of food aid to the companions of guardians in need. Rather than recognizing a 'right to be fed', this proposal offers a stopgap against, and partial remedy to, guardians who fail to appropriately feed their companions, whether due to neglect, ignorance, or some other reason. Thus, though practically and normatively viable, it cannot offer companions all that justice requires when it comes to food. To offer these things, we need to take a step further, and not only challenge neglect of animals' rights, but animals' exclusion from political decision-making processes.

Citizen Canine?

I have argued that companion animals have what looks like a positive right to food, but that this derives, in fact, from their negative rights. I have also argued that shifting the responsibility to feed these animals to guardians, even with state oversight, is insufficient. Not only does this fail to offer protection to *all* the companions who require it – it does not help those companions without an identifiable guardian, for instance – but it underestimates the level of change that would be necessary in a liberal democratic society to institute a right to be fed on the part of companion animals.

Why should we be concerned about those companions who lack guardians? They are part of a population that has a vulnerability to humans – and this vulnerability is of a normatively significant kind. They may have greater or lesser external dependencies – if they live on the streets, say, they may have very little, or if they live in a shelter, much more. But they certainly have internal dependencies, insofar as they belong to lineages that have developed (likely, been bred to have) certain genetic features. A dog does not become 'wild' just because she no longer lives under a human roof. Our collective responsibility for this vulnerability grounds a collective duty to aid these animals. For better or worse, the state is the body that has developed as the *de facto* discharger of these duties, and so it is the state that we should call upon to see that they are being done.

If companions – with guardians or otherwise – have a right to be fed, then we need to think seriously about the extension of *all* systems of food distribution to include them and their interests. So, for instance, hungry companions living with poverty-stricken but nonetheless loving and capable guardians could in emergencies be fed by the state, just as the state would feed the hungry children of poverty-stricken but nonetheless capable parents. More than this, however, the state would, as a matter of justice, have to implement social security measures to aid in the feeding of companions before they reached emergency levels, parallel to the kinds of institutions offered to humans in need. In most liberal democracies, this would mean some subset of the following: a system of companion benefit, analogous to child benefit, so that poor guardians are able to feed their companions; state-supported centres able to provide nutritious food to companions in need, analogous to (or, more likely, associated with) homeless shelters; funding for the feeding of companions at (newly) public institutions, analogous to the way that humans in need are fed at hospitals, schools, prisons, and the like; the provision of foods for companions at food banks, so that both human and nonhuman members of families in dire need can access nutritious food while retaining a degree of autonomy; the provision of emergency rations in areas affected by contamination, outages, natural disasters, and warfare; and so on.

There are all kinds of legal rights to food and drink in particular situations that are currently possessed by humans that could – and, likely, should – be extended to companions. There is no way that these can all be listed here, and the appropriate way to extend these cases (if at all) would need to be worked out in practice. For instance, in some jurisdictions, humans delayed on public transport have a legal right to be fed; in others, humans have a legal right to free water in bars and restaurants. If we are serious about protecting animals' right to be fed, then we would need to explore the extension of the various legislative protections and regulatory apparatus currently applied to the human food system to the companion food system. However, this would be because of the rights of the companions themselves, and not – as was discussed earlier – in the interests of further protecting the human food system or ensuring guardians' 'rights to know' about the foodstuffs they are purchasing. Companion foods as products would be entitled to the same kinds of protections offered to human foods; there could be injustices, for example, in excluding certain human foodstuffs from taxation while still taxing analogous foodstuffs intended for companion consumption, or in subsidizing human foodstuffs while failing to subsidize companion foodstuffs.

Companions' possession of a right to be fed thus calls for much more fundamental and far-reaching changes to the way we do and think about politics than simply instituting negative animal rights. This is true even if negative animal rights might have the more fundamental impact on the lives of many people and the practices of our societies – the end of animal farming as we know it, for instance. In a sense, negative rights call for *exclusion*. 'Stop doing that, leave them alone.' The possession of the positive right to be fed instead calls for *inclusion* in political thinking and political decision making. It calls on us to ask how we can include companion animals in our various food-related political decisions, and, crucially, political decision-making processes. This may sound like a big ask, and it is – but when we entered society with these animals, and when we cut off their possibility of an alternative future without us, we took on certain obligations. Many of these obligations are considerable.

If this is what is demanded of us, legally protected rights will not be enough (Cochrane 2020). When I say that all laws (in a certain domain) must consider companion animals' particular positive rights, what I am saying is that these animals need to be considered members of the community to which these laws need to be justified. They are members of the community who could legitimately (though metaphorically) say 'No – you have forgotten to take us into account.' Now, it makes little sense to say that the animals themselves can represent their interests like this. But this just speaks to the need to ensure that there are political procedures put in place to ensure that animals' voices are heard. It may be worth *literally* listening to animals' voices (Meijer 2019), but, at the very least, there need to be political decision-makers empowered and motivated to speak on animals' behalf. What we find, then, is that a close and serious consideration of companion animals' rights concerning food takes us much of the way towards the kind of political inclusion and political membership that is characteristic of the proposals in the political turn in animal ethics.

We lack the space to fully review the various practical proposals concerning the inclusion of companion animals in political decision-making procedures. In any case, such exploration would be beyond the scope of this chapter – existing proposals often stretch further than companions, and existing proposals concern much more than food policy. (Note that I am not committed to the claim that the kinds of arguments I am making extend *only* to companions or *only* to food policy.) But it is worth giving an indication of the direction that this kind of conclusion must take us.

First, companions' interests can be effectively represented in democratic decision-making processes only if we, in effect, extend the vote to them. It does not make sense to start literally sending them ballot papers, but there are proposals for how this could work in practice. For example, 'representatives' of animals could be selected by human deliberative assemblies. These representatives would then have a democratic mandate to speak on behalf of the animals, and would be subject to democratic recall if they failed to do so (Cochrane 2018, chap. 3). It is not hard to imagine that a model like this could ensure that animals' interests are represented in decisions about or related to food distribution. Individuals elected in this way could be empowered to identify decisions relevant to their constituency and then join votes in legislative bodies. There would need to be checks and balances to ensure that these representatives were acting in the interests of their constituents and that they were not becoming too powerful, but even the presence of a very small number of animal representatives of this sort would have the power to ensure that animals' interests were not forgotten in the drafting of laws that may affect their food supply.

Second, companions could be fully integrated into the political community with a recognition that they are citizens of the community in question – a proposal made famous by Donaldson and Kymlicka (2011). Part of reconceptualizing animals as citizens involves recognizing their 'equal right to communal resources and the social bases of well-being' (ibid., 142), and thus a right, like any other citizen, to be provided with food when in desperate need. This is a genuine positive right, owed to the animals because of their membership, and as such is not derived from a negative right. Donaldson and Kymlicka's vision is therefore stronger than the one developed so far this chapter – though represents, perhaps, an ideal to aim toward. It is about 'the commitment to constructing certain kinds of ongoing relationships that embody ideals of full membership and co-citizenship' (155), rather than a practical proposal to be instituted tomorrow. Recognizing the consequences of an extension of negative rights to animals opens the door to a fuller inclusion of companions in our political structures and systems. Some sense of citizenship represents, perhaps, the fullest understanding of that inclusion.

Why treat citizenship as an ideal at all? The extension of citizenship to animals, though a much stronger proposal than the one here developed, is motivated by the same kinds of arguments. Donaldson and Kymlicka argue that 'having brought [companions] into our society, and deprived them of other possible forms of existence (at least for the foreseeable future), we have a duty to include them in our social and political arrangements on fair terms. As such, they have rights of *membership*' (101, emphasis in the original). Rightfully or wrongly, we have made companions part of our community, and this is where they do and – likely – must remain. Human societies have, historically, done similar things with specific groups of humans, and it is now recognized that these humans (and their descendants) are entitled to be a part of the 'we' that makes up contemporary societies and these societies' common good. Given that my approach is grounded in a recognition of the importance of moral agency and responsibility – as well as my rejection of speciesism – it is natural that I must endorse animal citizenship (perhaps as defended by Donaldson and Kymlicka, but perhaps in some other form) as an ideal towards which to aim.

Animal democracy and animal citizenship are not likely to be instituted tomorrow. However, the negative rights I have defended could and should be realized very soon. These open the door to a more animal-inclusive politics – and it is a door that we should be prepared to step through.

Concluding Remarks

We have seen that our obligations concerning the feeding of animals depends upon whether we think of them as members of our family, members of other families, or something like our co-citizens; any of the three, depending on the circumstances, may be appropriate. We have a moral duty to feed our companions because of the nature of the relationship of vulnerability they have towards us, but this relationship of responsibility also grounds the kind of political responsibilities we have concerning them.

It is worth splitting what we should do as individuals from what we should do as societies. As individuals, we should feed our companions; to fail to do so is morally wrong. In providing food for them, we should be aware of issues of food justice, of our environmental obligations, and of what is good for the companions themselves. In particular, we should ensure that we do not *over*feed our companions. As individuals, it may often also be appropriate and morally commendable to feed other people's companions. This, however, comes with the important caveat that we must be wary of unduly interfering in the lives of others.

The same kinds of relationships of vulnerability and historical involvement in the shaping of (the lives of) companions writ large underpin some of what we *as societies* should be doing concerning the feeding of companions. Some of our societal obligations involve regulating and overseeing the pet food industry – indeed, we have good reasons to do this independently of any concerns about animal rights. However, the relationship of vulnerability between companions and humans, as well as the history of their relationships, results in something close to a right to be fed for companions. In practice, this results in a far greater need for political inclusion for companions. At the end of this chapter, I pointed towards some existing ideas of what this might mean in practice. Crucially, I suggested, it calls for the inclusion not only of those animals presently living as companions, but those animals living on the edges of human society but who have no home free from it – namely, 'strays', or other dogs, cats, and the like who do not live as companions in the typical sense.

This introduction of those animals who fail to be easily captured by the labels 'domesticated' and 'free-living', however, raises complications. Among animals, companions are the closest to us – both physically and socially. Meanwhile, free-living animals living well away from human habitation are a strong candidate for the farthest. Having explored the feeding of companions, it is time to begin to explore the multifarious relationships humans have with a range of animals that they do or could feed beyond those closest to them: their (animal) family members and (animal) co-citizens. Next, it is worth looking at what might seem like the most innocuous of practices: the feeding of garden birds. As will be seen, however, the complex normative obligations we have concerning these animals are demarcated by their status as not-quite-domesticated, not-quite-free, while the complex relationships that we have with such animals are contoured by their shifting status as garden pest or garden guest, nonhuman friend or nonhuman foe.

4

Animal Neighbours

We might think that little could be more innocuous than feeding the birds. Paradigm Western examples are provided by those who feed ducks in parks, pigeons in squares, and especially sparrows in gardens. Increasing presence and awareness of urban and suburban wildlife – as well as increasing openness towards them – has resulted in people choosing to feed a whole range of animals beyond birds they may once have considered a nuisance. Indeed, the idea that these practices warrant serious philosophical attention might seem bizarre. What could possibly be objectionable, it could be said, about purchasing birdseed and leaving it on a table for the finches who visit our gardens? It is perhaps this attitude that has resulted in the paucity of applied moral and political philosophy addressing the practice of feeding animals in this category – a category I call *animal neighbours*.

It is important to note that these animals *are* indeed in a distinct category. They are not companions, and nor are they animals created and used by humans for some other purpose. But they are not fully 'wild', either; though their lives are not controlled by humans, they live in human-shaped and human-inhabited environments, interacting with humans in ways that their more fully wild conspecifics do not and would not. Philosophers variously describe these animals as in the 'contact zone' between humans and animals (Palmer 2010), or else being 'liminal' animals – beings who are not quite members of human communities, and not quite members of their own 'communities' (Donaldson and Kymlicka 2011). For present purposes, however, these categories are too broad. Animals in Palmer's contact zone include all free-living animals who are affected by human activity; thus, while urban/garden wildlife is a paradigm example, a range of much 'more' wild animals are also captured. Indeed, in the 'Anthropocene', it may be hard to identify any animal who has not been affected by human activity (Keulartz 2016), and thus any animal who is not in Palmer's 'contact zone', to a degree. Donaldson and Kymlicka, meanwhile, stress the broadness of their category of liminal animals, including among them all non-domesticated animals who have in some sense adapted to life among – or minimally reliant upon – humans.

Despite first impressions, the feeding of these animals does raise a host of ethical questions, and these will be explored over the course of this chapter. First, we need to ask more about the relationship we have with our animal neighbours. Then, however, we can address four sets of normative puzzles. The first set concerns whether we have an obligation to feed our neighbours. The second is about the risk of creating a relationship of dependency. The third is about the risks imposed *on* the animals we feed *by* third parties. And the fourth concerns the impact *of* the animals we feed *on* third parties.

Before beginning in earnest, let us quickly note that food fed to animals in our gardens may be sourced in a way that is morally suspect. This will not be the focus of the present chapter. As explored in chapter 2, much use of animals in the production of food – even food to be fed to other animals – is unjust. This means that, for example, the feeding of (non-scavenged, non-cultivated) dairy cheese or (assuming they are, or may be, sentient) mealworms to birds, and the feeding of (non-scavenged, non-cultivated) meat-based sausages or pet foods to foxes, are practices that must be ceased. If we can source animal-based foods in a way that is ethically sound, then perhaps feeding them to our neighbours would not be wrong – indeed, as with our companions, we may have good reasons to believe that feeding them to our neighbours is less ethically questionable than eating them ourselves. At the very least, different ethical questions may be raised by the feeding of these non-vegan products to our animal neighbours than are raised by the feeding of them to ourselves, as has been discussed. And, again, some plant-based foods intended for animal consumption – as discussed in chapter 3 – may be produced in ways that conflict with our duties of food justice. Our duties concerning, for example, food workers' rights do not cease simply because the seeds we are purchasing are intended for garden birds rather than humans. Thus, many of the same duties that exist when it comes to the purchase of food for humans or companions apply when it comes to purchasing (or otherwise sourcing) food for animal neighbours. Though important, I take it that this normative ground has already been covered, and so I shall say no more about it here.

Conceptualizing Animal Neighbours

In this chapter, I am thus interested in what could reasonably be described as a subset of 'liminal animals', or animals in the 'contact zone'. Specifically, I am interested in those animal *neighbours* who live around our homes and spaces, interacting with us or the objects we have shaped to a greater or lesser degree (O'Connor 2013; cf. Acampora 2004). This category of animal neighbour – like its human equivalent – can be split further. Some neighbours are our friends: we enjoy seeing them, and they are welcomed into our spaces. Some we ignore: they get on with their lives, we get on with ours. Some neighbours, however, are a nuisance to us, or even adversaries – they are our animal foes. We feel that they interfere with our lives in troubling ways and make the spaces we share worse than they would otherwise be. For present purposes, the first and the last categories are the most interesting. Think of bird tables – we actively choose to feed our animal friends, leaving them foods we think they will appreciate. Meanwhile, we actively try to avoid feeding our enemies, even though these animals are often drawn to our spaces *precisely because* they hope to find food. Think of how much-maligned 'pests' and 'vermin' – gulls, rodents, pigeons, raccoons, canids, possums – will upturn our bins and take food from insecure stores.

These animal friends and foes are paradigm examples of *commensal* animals, at least as the term is sometimes used (O'Connor 2013).[1] That these animals should receive attention in a book about the ethics of animal diet is apt: the word *commensal* 'derives from "together at a table", reflecting the fact that commensalism generally conveys some feeding benefit' (ibid., 6). Many of those neighbours we ignore, however, are merely *synanthropes* – animals who share living space with us. Commensals are (paradigmatically) a category of synanthropes, and a category of significance when it comes to determining the morally relevant relationships we have with animals, and the normatively interesting feeding practices in which we – inadvertently or otherwise – participate.

Commensal animals, our animal neighbours, should be distinguished from a similar group with whom we have a very different kind of relationship (cf. Palmer 2003). These are animals who have been forced out of the spaces they once inhabited (typically, by human activity) or else have found that humans have entered and begun to heavily utilize the land on which they live. Their coexistence with humans – if they manage/are permitted to coexist – is thus uneasy, and very much forced upon them. Discussion of our relationship with such animals is worthwhile. One fertile line of enquiry, for example, is offered by ideas of wild animal property rights and wild animal sovereignty rights, which can mean that we wrong animals in using their space – or at least using it in ways that are harmful to the animals. However, this kind of enquiry is not my focus here. (In chapter 6, I will address wild animals under human control; in chapter 7, wild animals *not* under human control.) In the present chapter, I am concerned with a class of animals with whom we have a subtly but importantly different relationship. Commensal animals, in contrast to animals forced to live among us, have adapted at an individual or population level to life around human spaces, differing physiologically, ecologically, and ethologically from their more fully wild conspecifics. Just as domestic rats (including fancy rats and, for that matter, lab rats) differ from their riverbank-dwelling ancestors, so both differ from 'feral' (urban, street, sewer) rats, though all might belong to the species *Rattus norvegicus*. Commensal animals actively seek out human-inhabited spaces – their favoured habitat, ecologically speaking, is human-utilized land (for more, see O'Connor 2013).

Friends and Foes

We can differentiate animal friends and animal foes among animal neighbours based on feeding practices. It is meaningful to claim that differential practices of feeding are not merely consequences of these different relationships, but (at least partially) constitutive of them. As seen, biologists are prepared to define categories of animals around feeding relationships, and some philosophers of food have claimed that the very practice of feeding forms a distinctive kind of relationship. Indeed, they go further, claiming that food itself is best understood *qua* relationship(s). Ileana Szymanski (2016), for example, identifies food, rather than solely as assorted commodities, as a series of networks in which commodities are just one point. Eaters – who, she rightfully recognizes, can be human or nonhuman – make up separate points, while producers, norms, laws, and so forth all serve as more points. The act of eating (and, indeed, of refusing) food thus creates a relationship between eater and food, and in turn with she who provided the food, and through *her* a relationship with a range of other individuals, institutions, and norms. Put simply, feeding someone else creates and defines a relationship with them, and this relationship has normative significance.

Thus, for example, my purchase of birdseed draws me into a network of producers, growers, and retailers, but also societal norms (laws, customs, etc.) around the feeding of birds, and laws about the marketing and selling of foodstuffs. My *offering* of birdseed to a finch draws her into this network, and her choosing to consume the seeds consequently creates a particular relationship between she and I – a relationship wholly different to the one I have with other animal neighbours whom I do not choose to feed. Of course, in stressing the significance of these relationships, the corporeality of food should not be denied. No amount of sophisticated metaphysics can remove the fact that a slice of bacon is a piece of a pig's corpse.[2] But Szymanski's account (taken as representative of accounts like this) does not deny that. The commodities remain a part of the network, and, indeed, so too do the animals who are the source of the commodities.

The relationship I ground with a blackbird by feeding her seems like a friendship relationship. That is not to say that it is exactly like the relationships we have with human friends, which are surely (generally) richer in their content, involving a greater degree of reciprocity. When philosophers talk of interspecies friendship (see, e.g., Clark 2008; Jordan 2001; Townley 2010; 2017), the feeding of animal neighbours is not their focus. Rather than focussing on animal neighbours, these philosophers tend to talk primarily about companion animals. Pertinently, Cynthia Townley (2010, 50) refuses to commit to the possibility of garden birds being friends, even if humans feed them, and even if the birds deliberately interact with humans, as the relationship appears to lack some of what is required for us to conceive of it as friendship. Like Townley, I do not wish to commit to the claim that the animal neighbours we feed are friends in the sense with which these ethicists are concerned.[3] Nonetheless, I do contend that the term captures something important about what is going on between humans and the animal neighbours whom they feed.[4]

The opposite of the friend, of course, is the foe. Our animal foes are not simply those animals we choose not to feed, but those we actively try to *stop* feeding, whom we actively try to *stop* from taking things that were not meant for them. Those who raid our stores of pet food, birdseed, or vegetables; those who dig up our allotments or flower beds in search of roots; those who break into our bins in search of morsels – *these* are our animal foes. We erect barriers, or invest in more or less sophisticated scaring and locking devices, to *stop* these animals from eating in our spaces.

Obligations (and Permissions) to Feed

In chapter 3, I took it for granted that we were morally obliged to feed our companions, and sought an argument to justify that assumption. I will make no such assumption about our animal friends, and it should be obvious that the arguments used to justify the claim about our obligation to feed our companions could not be used to justify the claim about the obligation to feed our neighbours. We as individuals are not typically responsible for the plight of our animal neighbours, and nor, typically, are they dependent upon us as individuals for their wellbeing, food-related or otherwise.[5] It should be clear that we are, at least sometimes, *permitted* to feed our friends. Leaving seeds on a bird table for a family of sparrows, for example, seems like a paradigmatic example of a situation in which everyone benefits – the sparrows have access to nutritious, enjoyable food, and we take delight in watching them, or at least in knowing that they are not going hungry. (Of course, things are not always so simple.) It is hard to imagine an argument that rules out feeding *a priori* unless one has an implausible commitment to the value of wildness or the 'natural' (Palmer 2010, 78–84). Importantly, not only is such a commitment questionable in itself, but its applicability to the present case can be challenged, given that animal neighbours are not as wild or 'natural' as many more distant animals.

On a relational approach, the necessity of feeding animal friends is far from clear. Given that I have differentiated between the kinds of animal neighbours – animal friends, animal foes, and ignored neighbours – largely based upon food practices, it would be difficult to say that we are obliged to feed friends and not obliged to feed foes. Such a claim seems to simply endorse existing practices. Perhaps, then, if there is an obligation to feed animal friends, it must be an obligation related to their status as neighbours – a status they share with animal foes – and not their status as friends. Of course, the claim that we are obliged to feed all our animal neighbours is highly demanding, both because we have a great many neighbours with a great many food needs (made doubly difficult by our very clear obligation not to perpetuate the killing of animals for food), and because it is plausible that there are a great many animal neighbours we do not want to feed and encourage.

The analogy to human neighbours should be clear; trying to feed all our human neighbours for even an afternoon would be difficult and time-consuming, and, in many cases, we may not want to invite particular human neighbours to our dinner parties. This is despite the fact that, when it comes to human neighbours, we can talk about reciprocal relationships and the value of community-building in a way we cannot (easily) when it comes to our animal neighbours. Further, the demandingness of trying to feed all our animal neighbours would only increase in time, as more animals moved into our neighbourhoods to take advantage of feeding. These worries lead to problems. Conceptual issues arise when it comes to defining the neighbourhood; how do we draw a line around those animals who are a part of our neighbourhood, and those who are not? Ethical problems arise when it comes to justifying an obligation to feed neighbours but not other free-living animals; mere physical proximity, it is frequently claimed, has no moral significance.[6]

Justifying the claim that we are obliged to feed all our animal neighbours would be an uphill battle. Perhaps it would be more plausible to argue that we are obliged to feed friends, but not other neighbours, *even though* we are defining friendship based on feeding. Thus, while we would normally not be obliged to begin feeding our animal neighbours, we may be obliged to continue feeding them once we have entered a friendly relationship with them. We would not be obliged to continue feeding animal neighbours whom we have previously fed accidentally – insecure stores, morsels from bins, food meant for others – as these are not our friends. But this possibility, too, would require justification.

The philosophy of food contains an idea that may be useful in overcoming this impasse: *hospitality*. There have been several attempts in the food literature to resurrect the 'lost' virtue of hospitableness. Elizabeth Telfer (1995; 1996, chap. 5), for instance, offers an in-depth analysis of hospitableness as an 'optional virtue' – or imperfect duty[7] – intimately and inextricably tied to food.[8] Let us attend a little more closely to Telfer's understanding.

Hospitableness is tied to hospitality, which is 'the giving of food, drink, and sometimes accommodation to people who are not regular members of a household' (Telfer 1996, 83). If we allow a sufficiently flexible account of 'people', the feeding of animal neighbours is thus a paradigm example of hospitality. Telfer offers three kinds of hospitableness: (1) Hospitableness tied to one's role within a social circle (including hospitableness to family members, colleagues, and so forth); (2) 'Good-Samaritan hospitableness' – which responds to need in another; and (3) Hospitableness towards one's friends (ibid., 90–5). The feeding of animal neighbours falls most comfortably into the third category. It could, in certain circumstances, fall under the second, but the fact is that humans *choose* to feed certain neighbours and not others. This is not out of a belief that those fed are in need, in contrast to those not fed; it is out of a desire to spend time around and/or support those fed, in contrast to those not fed.

Hospitality, Telfer argues, is particularly appropriate as a means of entertaining friends, since the sharing of food – and, as in paradigmatic cases of the feeding of animal neighbours, the invitation and welcoming into one's own space – has a certain kind of intimacy that may be less appropriate for non-friends. People entertain friends because 'liking and affection are inherent in friendship', and 'the liking produces a wish for the friends' company (as distinct from company in general), the affection a desire to please them' (1996, 93). Thus, Telfer's ideas of hospitality/hospitableness fit well with how I have characterized the relationship between humans and those animal neighbours that they feed. Telfer's account thus serves not only as a viable lens for understanding ethical obligations but also as an argument that the relationships we have with the animal neighbours we do choose to feed are, at least minimally, friend-like.[9]

We might say that hospitality is not a viable lens for talking about the feeding of animal friends because we do it – generally – for our own pleasure. But there is surely a concern for the animal friends themselves. We want a food that they will like; this is not just so that they keep coming back, but because we want them to be happy. A gardener who plants *Buddleja* plants purely because he wants to catch butterflies for his collection, or because he wants butterflies to pollinate his other plants, is not being hospitable to butterflies, but nor would we want to call the butterflies his friends. They are mere objects or tools for him. But someone who leaves birdseed with the intention of watching birds, showing genuine concern for the happiness of the birds themselves, is indeed a hospitable person, just as, in Telfer's example, someone who invites guests to alleviate her own loneliness can be praised for being hospitable, despite her initially self-concerned motivation (1996, 88).

As indicated by her characterization of hospitableness as an imperfect duty, Telfer does not believe that we are obliged to feed everyone.[10] There are strong limits on hospitableness; the refusal of hospitality may sometimes be appropriate, either at a first encounter or after a guest has overstayed her welcome (Boisvert 2014a, 1184). Thus, it is not unreasonable to deny hospitality to those animals we do not want to be around – our animal foes. If we do not wish to share our food and spaces with certain animal neighbours, those neighbours have no right to them. And, provided we respect their negative rights, these animal neighbours can be kept away. Barriers difficult for them to cross or smells that they find unpleasant can be deployed to protect our stores, our refuse, or our vegetable patches, even while traps and poisons cannot. We can even deny hospitality to (former?) animal friends who now (figuratively or literally) demand too much. The extension of benefit on one occasion is no commitment to continued extension indefinitely, *even if* we may owe certain obligations after an invitation has been offered (as will be explored below). While it is good to be hospitable, and it is the appropriate way for us to treat our animal friends, it cannot and should not bind us to a duty to provide for animals indefinitely. On this point, it is perhaps worth appropriating some of Immanuel Kant's words on hospitality in the political domain: 'If it can be done without destroying him, he can be turned away; but as long as he behaves peaceably he cannot be treated as an enemy' (2003, §358). As a rule for the treatment of animal neighbours, this has much in its favour.

But this account may leave an unpleasant taste in the mouth. Is this not a cover for bigotry? Consider someone who chooses only to feed/befriend her white neighbours, while treating her non-white neighbours in a non-friendly (though respectful) manner. Is there a difference between this person, and the person who feeds the birds but not the squirrels?

Two distinctions can be made. First, the non-extension of hospitableness in the human case might be particularly egregious when it falls foul of hospitableness tied to one's role within a social circle (Telfer 1996, 90–1). If I invite everyone on my street to a party *but for* my Polish neighbours, this seems inhospitable, as I have artificially divided the relevant existing social circle – but this relates to a different kind of hospitableness/hospitality than is at stake in the animal neighbour case, which is a case of (if you like) *creating* a social circle. Second, we can distinguish between traits on which it is more and less reasonable to make decisions about the extension of hospitality. It would be reasonable to not extend hospitality towards a (potential) friend because he is particularly unhygienic, or has particularly messy eating (or toileting!) habits, or because he eats more than his share, or because he bullies other guests. When it comes to the deliberate exclusion of some animal neighbours, one needs to ask what the motivation of said exclusion is. It is one thing to not extend hospitality to dirty, aggressive, or gluttonous birds – assuming that one's assessment of dirtiness, aggression, or gluttony is not based upon insidious stereotyping. It is quite another to not extend hospitality to drab coal tits because one wants a garden full of flashy long-tailed tits. *Some* non-extensions of hospitality may have a loose analogy in racist inter-human decision-making. Others will not. *Some* may be unvirtuous; others will not be.

We thus have the theoretical tools necessary for understanding and demarcating our obligations concerning the feeding of our animal neighbours. It is now time to turn to the other side of the question, and – drawing upon these tools – explore potential problems raised by feeding in certain cases.

Feeding and Dependency

In paradigm cases,[11] we can sharply demarcate our animal family and our animal friends by their degree of dependency upon us. Our animal family members require us to feed them, due in a large part to the extent to which we have shaped their lives (both before and after their birth) and the control we exert over their activities. Animal friends, on the other hand – excluding the occasions in which their plight calls for us to exercise 'Good-Samaritan hospitableness' (Telfer 1996, 91–3) – have the ability and the freedom to live freely of both us and our personal influence. They are free to choose to sever the relationship we have with them, free to leave and never return, free to refuse some or all of the food we offer them, and so on. Put simply, if they want to, they don't have to come into our garden. Equally, as discussed, we have a similar degree of freedom in return. We do not wrong the birds we once fed in moving to a new house, though we surely *would* wrong a companion were we to leave her in our old house when we move away. Similarly, if times are hard, we can simply stop our purchase of birdseed, but to stop feeding a companion, we would have to make alternative arrangements to ensure that the companion could eat elsewhere.

These moral differences map onto differences in dependency between our animal family and our animal friends. Importantly, this means that to retain the moral freedom to withdraw from the friend relationship with animals, we must keep the relationship sufficiently distant, so that our animal friends never become dependent upon us (again, excluding rare emergencies). If we fail to keep relationships sufficiently distant, then we lose the ethical permission to withdraw from the relationship; we appear to have transformed our animal friend into an animal family member, morally speaking.

This kind of transformation is not unique to human-animal relationships. Imagine a lonely woman has a friendly, neighbourly relationship with a child who lives nearby. The woman may offer the child snacks or a drink when he visits, and he may accept, staying with her a short time. She may also help him out in cases of emergency; for example, his parent has faced an emergency and left the house, leaving him hungry and alone. The woman's relationship with the child is, morally speaking, analogous to the relationship she has with birds she feeds, and the child would have little claim against her were she to choose to leave. Imagine, though, that the emergencies faced by the child become increasingly common, until the woman finds that she is caring for him a lot of the time; she provides most of his food, and he spends most of his days at her house. In this case – and though a clear line might be impossible to draw – it seems that her relationship with the child is now more analogous to her relationship with her dog than with 'her' birds.

This is not to say that there is necessarily anything *wrong* with blurring the lines between animal friends and animal family. Take Len Howard, a twentieth-century British ornithologist and musician who literally opened her house to birds. Howard would feed the birds twice a day, but, for the most part, they would come and go, and she never allowed them to become dependent. There was one exception: a lame bird to whom she tended for a year before release (Meijer, personal correspondence). It would be fair to say that though Howard's relationship with the birds came close to – and, in one case, crossed – the hazy line between 'animal friend' and 'animal family', she did not wrong the birds by having such a close relationship with them. On the other hand, were she to abandon the lame bird, or release her before she could fend for herself, she would have wronged the bird. Given, however, that Howard had the ability and will to offer the necessary care to the bird in question, the bird was not wronged by her move from friend to family member.

Howard's case, to be clear, is exceptional, and we should generally take precautions not to turn our animal friends into animal family; not only does this place considerable demands upon us, but we risk harming animals who may not benefit from a closer relationship with us – who, for example, would not welcome a life mostly cooped up inside a house. We also risk adversely affecting third parties, such as the animal's dependent young.[12]

Dependency relationships can also be created *as collectives*, and this complicates matters. Palmer (2003, 73–5) draws attention to large numbers of pigeons who lived in Trafalgar Square and the debates around their feeding – including the likely outcome of pigeon death and suffering if legislation limited the feeding of these animals. It is a human collective (the precise makeup of which might be difficult to pin down) that is responsible for the dependency of these pigeons on humans (it was the feeding by people that allowed so many pigeons to survive). And this collective responsibility means that, though these pigeons had not become family to any one person, a human group does have a significant responsibility towards them, just as a human group continues to have (I have already argued) a significant responsibility towards companion animals who are living in a 'feral' state. The pigeons thus have a justice claim – though pinpointing precisely who or what owes them something (for example, continued feeding) is going to be tricky.

This is not to say that the banning of the feeding of pigeons on Trafalgar Square was necessarily unjust, but it does mean that the pigeons would be the victim of an injustice if humans were to simply wash their hands of the birds and leave them to suffer and die. A range of possible alternatives to simply banning pigeon feeding could be explored. For example, the use of contraception to thin numbers in conjunction with a gradual cessation of feeding; the provision of feeding at some other location; or even the removal of pigeons to a more domesticated setting, analogous to an animal shelter taking in 'feral' companions, may be plausible. Real-world strategies of this sort have seen success in reducing pigeon numbers in a respectful way, even while preserving the relationships of feeding valued by some people.

It is worth noting that dependency can be created without humans (individually or collectively) being morally responsible for that creation, and it is unreasonable to demand that humans feed and care for animals who have become dependent on them through no fault of humans. There is a clear difference between the bird-lover who, through her activities, fosters a close relationship with a well-loved, now-dependent dove and a gardener whose vegetable patch, through a series of unfortunate circumstances, becomes the only source of food available to a group of rabbits. These rabbits are genuinely dependent on the gardener, but it seems strange to say that the gardener is obliged to care for the rabbits just as the bird-lover is obliged to care for the dove, not least because the gardener is more likely to view the rabbits as *enemies* than as friends or family.

Provided that reasonable steps are taken to ensure that individual animals do not become dependent – and the gardener, presumably, has taken steps to keep her vegetables out of rabbit mouths – it is fair to say that individual humans lack moral responsibility for the development of a dependency relationship, and thus the relationship has a different character than if the human agent were responsible. Indeed, the very act of feeding in this case seems to differ, morally speaking. Were the gardener to feed the rabbits, it would surely be an example of Telfer's 'Good-Samaritan hospitableness' (1996, 91–3), and not the hospitableness towards friends that first motivated the bird-lover to feed the dove. (Framing the gardener's feeding of the rabbits as a 'Good Samaritan' act is particularly apt; in Christian teaching, the Good Samaritan is praiseworthy not least for helping a Judean, given that Samaritans and Judeans had a hostile relationship.) It is of course the case that it would be *good* of the gardener to feed and care for the rabbits in this scenario, but to say that the gardener would *wrong* the rabbits by failing to do so appears to mischaracterize the nature of her relationship with them, and the difference between this relationship and the human-dove relationship. While the latter is very close to a relationship between a companion and her guardian, the former has a different character; someone encouraging the gardener to feed the rabbits might appeal to mercy, compassion, and sympathy, but appealing to the rabbits' rights and gardener's (perfect) duties appears to miss the mark.

Risk to Friends

The imposition of risk of bodily harm – even if it does not result in any *actual* bodily harm – can be a serious wrong. If I cannot recall whether a particular foodstuff is safe for the hedgehogs I feed, I wrong the hedgehogs by choosing to leave it out for them anyway, even if, as it turns out, that foodstuff is perfectly safe for hedgehogs. (A recurring problem in the UK is that people mistakenly feed hedgehogs milk, which is very bad for them.) Among other things, this would be a failure of hospitality. A more interesting sense in which our feeding of animal neighbours puts them at risk, however, is the risk they face of predation. Passerines entering a suburban garden, for example, are put at risk of becoming food – both food for other animal neighbours, such as passing raptors, and for companions, such as free-roaming cats. Our obligations towards our animal friends concerning these potential predators is worth exploring.

We can distinguish between a set of moral reasons we may have for taking steps to limit the risk of predation – in line with my earlier discussions, I will frame these as (imperfect) duties of hospitality – and a set of rights-based reasons we may have for taking such steps. We can also distinguish between the threats posed to animal neighbours by other non-domesticated animals (e.g., raptors) and domestic animals (e.g., cats).

The link between hospitality and safety should be clear. In Telfer's words, key to hospitality is a host accepting 'responsibility for the overall welfare' of her guests; indeed, 'traditionally the most important responsibility of all was for the guest's safety – hospitality was a kind of sanctuary, and the host was thought of as having undertaken a solemn obligation to make sure no harm came to his guest while under his roof' (1996, 83). Thus, householders have a moral responsibility to take steps to create a safe space for their animal friends; to invite them into spaces in which they have relative safety from predators. This duty need not be as onerous as it might sound. For example, the bird table itself is, at least in part, a means of ensuring safety; birds feeding on a table are away from the ground, giving them precious seconds to escape from terrestrial predators.

Telfer's talk of a 'solemn obligation' (1996, 83) need not worry us, either. The reason I say this is twofold; first, Telfer here addresses the traditional sense of hospitality, as used in the classical world. The appropriate conceptualization of hospitality for a modern ethics is somewhat more muted. However, this does not make it impotent. Let me give a relevant example. It would be a failure of hospitality to feed human guests off dirty plates – that would not be to take responsibility of the welfare of the guests. Equally, it would be a failure of hospitality to feed our bird guests using a dirty seed dispenser. This is a very real problem in suburban gardens, with feeding stations becoming hives for transmittable bird diseases. Second, there are degrees of hospitality. Consider the different degrees of a hospitable relationship afforded by the walker who feeds ducks in a park, the householder who feeds birds in her garden, and Len Howard, who literally welcomed birds into her house. The walker merely shares food and company with the ducks. The householder invites the birds into a space over which she has a level of (incomplete!) control. Howard, meanwhile, shares food, company, and space – in all three cases, intimately. Consequently, the walker has very limited special duties concerning the safety of the ducks, and though the householder possesses some greater level of said duties, they are by no means overwhelmingly onerous. Only Howard approaches a 'solemn obligation' when it comes to her special duties concerning her animal friends – it is fair to say that she does have strong obligations concerning these birds' safety when they enter her house, just as we have a strong obligation for the safety of our human friends while they spend time with us in our house at our invitation.

I conclude that we have good moral reasons to extend *a degree of* protection to our animal friends, at least while we are welcoming them into our company/space. The nature of this will vary, depending upon the closeness of the relationship we have with them. Two caveats are worth making explicit. First, this moral duty extends only to our animal friends, as it is they to whom we extend hospitality. I have no special duties – or, at least, none derived from hospitality – concerning the protection of my animal foes and those animal neighbours with whom I share mutual disinterest.[13] Second, the moral duty extends to our neighbours only while we are extending hospitality, and – crucially – only while our animal neighbours freely accept this hospitality. We overstep our role in the relationship when we force our 'hospitality' upon a neighbour; for instance, when we take steps to control the movements of our neighbours in the name of their protection. Note that this opposition to controlling the lives of neighbours is not a rights-based claim relying upon the nature of our neighbours' interests (though one could, of course, make such a claim); instead, it is a claim about the nature of our relationship with them.[14] A different kind of relationship might permit (or even require) much greater control over the animal's life – as explored in the previous chapter.

Justice-based considerations enter, however, when it comes to the risk faced by animal neighbours from domesticated animals, especially cats. When an animal – friend, foe, or mere neighbour – is killed by a domestic cat, that animal's blood is on our hands, both as individual guardians and as members of the mixed society, just as the blood would be on the hands of the parents and of society if a young child were to kill these animals. Though cats and young children are not moral agents, and are thus unable to violate rights themselves (Regan 1984, 285), this is not the end of the story. Instead, moral agents – such as paradigmatic human adults – can be morally responsible to a greater or lesser degree for the harm (causally) inflicted by non-agents (Milburn 2015b; see also Jamieson 2008, 186–7; Palmer 2010). The careless guardian is not *as* responsible for the death of birds as she would be if she had deliberately encouraged her cat to hunt; even then, she would not be *as* responsible for the birds' deaths as had she shot them herself. Nonetheless, she would have a degree of moral responsibility, and thus could be held responsible, to a degree, for the violation of the birds' rights. The same can be applied, *mutatis mutandis*, to society writ large; as explored in the previous chapter, these cats are here because of us, and thus the collective 'we' has some responsibility for the harms they perpetrate, and thus, to a degree, 'we' are responsible for the right-violating harms inflicted upon our neighbours by them. Thus, every one of us has *some* reason to intervene to protect (say) birds from domestic cats beyond the reason (if any – see chapter 7) we have to protect birds from wild cats. In both cases, there will be harm to birds – but in one case, we (as individuals) belong to a collective 'we' that shares in some responsibility for the harm.

As individuals, but especially as collectives (and, here, I am thinking of states), we have obligations to defend the victims of rights violations. When we see cats threatening our animal neighbours, we should recognize this as an unjust failure of individuals and the state to adequately protect the rights of animals, and thus intervene if we are able to do so without putting ourselves at risk or considerable inconvenience. To be clear, it is not the cats who are behaving inappropriately, here. Cats are not moral agents, and are thus innocent. Instead, the cats' guardians, or the state/society that has created an environment in which cats can pose such a threat, is to blame. What this means, of course, is that it is not the relationship we (understood broadly) have with our animal neighbours that grounds this duty of justice, but the relationship we (understood broadly) have with our companions. The fact that cats pose a threat to animal neighbours is hardly a new observation (see Marra and Santella 2016), but what does have to be recognized – and seldom is – is that it is *individual* birds (and rodents, lizards, and other neighbours) who are the victims of injustice, and that the metaphorical blood of these animals is on the hands of humans, *not* the paws of cats.[15]

Framing the injustice of birds' deaths due to the actions of cats around the question of moral agency has a distinct advantage over an alternative approach. Donaldson and Kymlicka argue that cats, as members of our community, must be regulated to protect those animals they would harm: 'part of our responsibility as members of a mixed human-animal society is to impose regulation on members who are unable to self-regulate when it comes to respecting the basic liberties of others' (2011, 150). Donaldson and Kymlicka do not extend this duty to encompass responsibility over those liminal animals who live among but not with us; cat citizens must be stopped from hunting denizen pigeons, but we need not stop denizen peregrines from hunting the pigeons. However, the citizen/denizen distinction is not able to do the work it needs to for this to hold, for the simple reason that there are some denizen animals we should be preventing from killing other denizens; namely, those animal companions who are (appropriately thought to be) denizens by virtue of the fact that their *guardians* are (appropriately thought to be) denizens (Milburn 2016c, 201–2). It would be bizarre to suggest that (for instance) a migrant worker arriving in a state is a non-citizen but that the non-human members of her family become citizens upon arrival; much more reasonable is to say that the companion is a denizen just as her guardian is.[16] But it seems arbitrary to claim that the bird killed by the denizen companion cat is the victim of no injustice while the bird killed by a citizen companion cat is a victim of injustice. We certainly would not make the analogous claim in the human case – animals (and, indeed, humans) are as entitled to protection from the aggressive human denizen as they are to protection from the aggressive human citizen. (This includes non-agent denizen humans, like denizen *children*.) To preserve the claim that liminal animals warrant protection from both citizen and denizen companions *without* claiming that liminal animals are entitled to protection from each other, it is prudent to look to moral agency rather than any citizen/denizen divide.

Putting Others at Risk

The reverse of the risk that third parties may impose upon our animal friends is the risk that our animal friends may impose upon third parties. Given that my examples have concerned the likes of sparrows and hedgehogs, this may sound like a non-starter. However, humans feed a range of animals who may impose risks upon others. Foxes, for instance, pose a risk to small mammals, and widely hyped cases of foxes attacking companions and supposedly even children have been seized upon by detractors (O'Connor 2013, 70–1). And, of course, small birds can harm morally significant others if we entertain broader accounts of 'harm' or 'morally significant others'. Starlings, where they are found, are the bane of house- and car-proud humans for their propensity to cover windows and vehicles with faeces, while less lyrical birds – grackles, crows, ring-necked parakeets – can be framed as auditory nuisances. Many birds are also, of course, predatory. On an animal rights account, the fact that animals we invite into our spaces may kill and eat other animal neighbours cannot be dismissed. I do not want to get drawn too far into the question of whether worms eaten by blackbirds are wronged – invertebrates, as noted in chapter 2, are already tricky on animal rights accounts. But many householders are excited at the prospect of inviting and welcoming birds (and other animals) who prey upon birds, rodents, or reptiles.

Let us look first at the latter issue. The risk imposed by our animal friends on prey animals takes on a different character depending on whether we have a level of responsibility for the prey animals. It should be obvious that we are failing in our responsibility to our animal companions if we impose risk upon them by inviting their predators into our spaces. Choosing to feed foxes in a garden in which companion chickens or rabbits live is to choose to impose an unjust risk on our companions. Compare: while it may be permissible or even (in the right circumstances) commendable to break bread with a hungry but criminally violent individual in our home, we have special obligations towards our dependent children, meaning that it would be wrong of us to expose them to the risk of attack from such an individual. We should thus not invite this person into a home we share with our children, or at least ensure that steps are taken to be certain that the individual posed no risk to the children.

We also, as already discussed, have certain (though much more limited) moral duties of protection when it comes to those animal neighbours to whom we extend hospitality. If I am extending hospitality to A, I thus have a *prima facie* duty not to extend hospitality to B if there is reason to believe that B will visit harm upon A while both are my guests; most obviously, this means not actively extending hospitality to both predator and prey concurrently and contiguously. While it might be commendable to bring (temporary) peace between enemies by having them eat together, this works only if such a move *will* bring peace. A range of values – honour, respect for a host, and so on – might lead even mortal enemies to peacefully eat together at the home of a neutral party, but peace between nonhuman predators and prey is less likely. We do wrong if we leave food for rabbits in the same place that we leave food for foxes – and the rabbits are the party whom we have wronged, as we have failed in our duties of hospitality to them. While our duties to protect our animal friends are, as already explored, quite minimal, they surely extend sufficiently far that we are obliged not to *actively encourage and support* the presence of those who would inflict severe harms upon them.

These kinds of duties of protection do not extend to those of our animal neighbours to whom we do *not* extend hospitality – especially not our animal foes. While (as with the previous example of the vegetable patch and the starving rabbits) we may choose to extend compassion and mercy to our animal foes and offer them some level of hospitality, and thus protection, we are far from obliged to do so. We do not gain a special responsibility to protect those – human or animal – who enter our spaces uninvited; it is not clear that this creates a new morally salient relationship grounding any obligation of protection.

This exploration of hospitality is not all that might be said about the harm our animal friends inflict upon our other animal neighbours, however. It could be said that the latter have a claim upon us because their negative rights have been violated, and thus that there is an issue of justice at stake. Let us explore this by way of example. Imagine I have a large garden sometimes visited by sparrows, but that I have a fondness for sparrowhawks, and so attract the latter to my garden with food.[17] Let us also assume – as I take to be true – that sparrows generally possess interests sufficient to ground negative rights against being made to suffer and against being killed. In this scenario, it is safe to imagine that the sparrowhawks, as well as taking the food that I offer them, will inflict suffering upon and kill the sparrows. While these sparrows have no claim against the sparrowhawks – who are not moral agents – they arguably have a claim against me. I cannot kill the sparrows or make them suffer. That would be unjust. I must also take steps to prevent my companion cat from killing them or making them suffer – if I fail to do so, I will be responsible for the violation of their negative rights (to a degree). The sparrowhawks are causally responsible for the harm to the sparrows, but, as I am causally and morally responsible for the presence of the sparrowhawks, it could be argued that I am causally and morally responsible *to a degree* for the harm to the sparrows (though to less of a degree than I am for harm inflicted by my companion cat), and thus blameworthy for a rights violation.

This is an important challenge. I would presumably be *more* responsible for the harms to the sparrow (morally, causally) in a similar case in which the sparrowhawks *and* sparrows were my friends than in the case in which the sparrows and I share mutual disinterest. In the former case, I would have moral (and causal) responsibility for the presence of the sparrowhawks *and* the sparrows, and thus a greater degree of moral responsibility for the resulting harms. This would mean that not only was I failing in my moral responsibility to my sparrow friends but that they had a claim *of justice* on me. Conversely, my responsibility for harms to the sparrows would be lessened, or even eliminated, if the sparrows were my animal foes, and I had consequently taken steps to limit their presence. It is hard to imagine that beings I am actively trying to keep out of my space have a claim of justice on me because of harms visited upon them by non-agents while they are in my space. I cannot visit violence on them, but I surely owe them no special obligations of protection. (As already noted, I may have general obligations to protect them from injustices, but we are here asking whether there are any rights violations/injustices taking place.)

But what if the sparrows in the imagined example are neither my foes nor my friends? (For many people, most animal neighbours will fall into this category.) If the challenge is well-grounded, then it perhaps follows that we have a duty not to befriend *any* animals who may enact violence upon animal neighbours, excluding any animal neighbours whom we actively discourage from entering our spaces. This would mean that if we *did* want to feed any carnivorous or omnivorous neighbours, we would have to take steps to ensure that prey animals were not present in our spaces. This could be highly demanding. It is worth exploring whether I can avoid this conclusion without abandoning my claims about degrees of moral responsibility.

One strategy would suggest that the degree to which an agent is morally responsible for harm would have to be of some sufficient level (with 'sufficient' here meaning something more demanding than 'greater than zero') for the harmed subject to have a claim of justice against the agent. If A and B are subjects of justice but not moral agents (i.e., A and B are moral patients), and B is causally responsible for significant harm to A, and C is to some degree *morally* responsible for the harm B causes to A, we would need to ask whether C is *sufficiently* morally responsible for the harm to correctly say that A has a claim of justice against C. This seems to be importantly correct. Let us imagine two birds are killed in the street behind Smith's house. One is killed by Smith's cat, and the second is killed by a cat Smith shooed out of her yard. In both cases, Smith seems to be – to a degree – responsible for the harms inflicted on the birds. But only in the case of the bird killed by *Smith's* cat does it seem that there is plausibly a claim of justice against Smith. While Smith is morally responsible for shooing the cat, and while it is foreseeable that the cat could go on to attempt to kill any bird that happened to be in the street behind the yard, and while Smith *could* have enticed the cat into her house and provided delicious (plant-based) food until the cat would have little desire to chase birds, it seems excessive to say that these facts amount to a claim that Smith is responsible for an injustice visited on the bird. (*Someone* is likely responsible, however, and it would be very easy to make the case that Jones, walking along the back street, has good reason to intervene in any cat-on-bird violence she witnesses.)

Applying this line of thought to the previous case of the sparrowhawks and sparrows in my garden, it could be said that I *am* minimally morally responsible for the death of the sparrows killed by the sparrowhawks, but that this level of responsibility is not sufficient to say that the sparrows have a claim against me. This would still leave open the possibility, of course, that sparrows to whom I have extended hospitality would have a claim against me concerning sparrowhawk attacks (even beyond the already-discussed duties of hospitality), as I have *greater* moral responsibility in this case, and so I am more likely to have 'sufficient' levels of responsibility for the sparrows to have a claim against me.

There is also, I think, something to be said about the importance of intentions. Regardless of the outcome, there seems to be a clear injustice in the case in which I entice sparrowhawks to my garden with the expectation and hope that they will kill sparrows in contrast to the case in which I befriend sparrowhawks but regret harm they cause to resident sparrows. I seem to have greater responsibility for the harm sparrowhawks cause in the latter case than the former. If I try to bring something about, I am more responsible for that thing occurring than if it comes about through my actions foreseeably but not deliberately. I am less responsible still if I do not intend for it to come about, and could not reasonably be expected to foresee it coming about.

Thus, I am inclined to think that, in many cases, we should not worry too much about the harm that our animal friends may be causing. We are not responsible for their actions, and, providing we do not negligently or maliciously contribute to the harm they cause (say, by putting some third party in harm's way), others are unlikely to have claims of justice against us concerning their actions.

Human and Animal Neighbours

In closing this discussion, it is worth looking towards the negative impact that our animal friends may have on our *human* neighbours. A whole book could be written about this, so I can only gesture towards answers here. Much of what I have said above continues to apply; in many cases, it would be a stretch for human neighbours to say they have a claim of justice against us because starlings had dirtied their car. However, there are morally salient features of the human-neighbour/human-neighbour relationship lacking in the human/animal-friend relationship. Typically, our human neighbours (and, in an ideal world, if not always the real world, their companions) will have particular relationships of co-citizenship, co-nationality, and co-community-membership with us. Despite the strong *personal* relationships we can have with our animal friends, we cannot (easily?) have these *political* relationships with them. This is recognized by many political approaches to animal rights (e.g., Donaldson and Kymlicka 2011; Smith 2012; Valentini 2014), which focus on companions and other animals more clearly under human control as members of a mixed human/animal community, conceptualizing our animal neighbours – if at all – differently.

This has two key consequences. First, we may have important reasons to take the interests, desires, conveniences, and so on of these people more seriously (all else equal) than we take those of our animal neighbours, perhaps even including our animal friends. While our human neighbours' irrational prejudices – and, let us be clear, it is irrational prejudices that often motivate people's distaste for (sub)urban wildlife – should hold little sway over our decision-making, any *legitimate* concerns they have about our animal friends doing harm to them, their property, or *their* animal friends should be considered in our moral decision-making. This should not be a controversial claim; I have a responsibility to consider the negative impact that my hobbies, activities, and interests will have on members of my community, and to moderate my behaviour if necessary. What this means is that animal lovers cannot simply ignore their human neighbours in favour of their animal neighbours, even if the extent of the 'harm' that their animal friends cause their human neighbours is minimal.

Second, however, it is these co-citizens who legitimately contribute to political decision-making, from the local to the transnational level. And, crucially, governments of all levels might legitimately make decisions that affect our animal neighbours and the relationships we can have with them. There is nothing in principle unjust about communities deciding to regulate certain forms of interaction between humans and their animal neighbours; for instance, as already indicated, there is nothing *inherently* unjust about the bylaw banning the feeding of pigeons in Trafalgar Square.

Concluding Remarks

The entanglements we have with the animals we choose to feed (and those we do not want to feed) in our gardens, parks, and the like can be thought through using the language of hospitality. We choose to extend hospitality to some animals, and we can think of them as friends. This brings with it certain obligations – but they are not particularly onerous obligations. Other animals, however, we do not extend hospitality to, and we seek to stop them eating our foods. These are our animal foes. That is fine; we have no obligation to feed those animals who live around us, excluding (supererogatory) 'Good Samaritan' duties to those in dire need. In some cases, though, our duties towards our animal neighbours seem to step beyond questions of hospitality to duties of justice. The most pertinent case is the predation of birds by cats. While the cats cannot violate rights, the birds do have a claim of justice against *us*.

At various points in this chapter, we have pointed towards the complex situations in which large numbers of animals are vulnerable to our collective decisions – such as the pigeons, vulnerable to our decision to ban feeding in Trafalgar Square. In the next chapter, we will explore a particularly stark and troubling example of this sort of case. In so doing, we will also continue our journey from those animals closest to us towards those further away. Having explored (sub)urban neighbours, we now turn to a particular kind of rural neighbour – those animal neighbours who live on, and around, our farms.

5

Animal Thieves

In the previous chapter, my attention was given over to those urban and suburban animal neighbours with whom we can forge personal or collective relationships of friendship or enmity. In the present chapter, I turn to a particular kind of animal neighbour far more associated with rural environments; namely, those animal 'thieves' who live among and around arable farms. This group includes, but is not limited to, a variety of rodents, birds, reptiles, and invertebrates (though, as earlier, I will not focus on the latter because of the additional puzzles they raise). Some of these animals are associated strongly with arable farmland (think of *harvest* mice or *corn*crakes), while some are opportunists equally at home in cultivated and uncultivated spaces (like rats). This group has caused controversy for animal ethicists, and is worth attending to if for that reason alone.

The controversy arises because animal thieves are drawn to arable farms by the food we grow for ourselves (and other animals), and our harvesting practices consequently present a mortal threat to them. This threat is at the core of the anti-vegan argument pushed by those we can call 'burger vegans' – though the term may be unfamiliar, the challenge is common in both academic and popular discourse.[1] In a sentence, burger vegans argue that the vegan's moral commitment to minimizing animal harm, paradoxically, should encourage them to eat some meat. Thus, burger vegans do not deny the moral significance of harm to animals. But they argue that the harms caused by arable agriculture are sufficiently high that animals' advocates should favour a diet containing some meat rather than a vegan diet: that is, that a diet containing meat is actually less harmful than a fully vegan diet. Now, as I will argue, the burger vegan's argument fails if it seeks to justify the eating of a conventional meat-based diet rather than a vegan diet. Indeed, it falls flat even when it is used to justify the eating of beef from grass-fed cattle, as favoured by some of its proponents.

But the fact that burger vegans have failed to show us that we should start eating beef does not mean that harm to animal thieves has gone away. The best response to the burger vegan, I contend, is to develop forms of arable agriculture that are significantly less harmful to field animals. Some animal ethicists have started to sketch what this may look like: clever deterrence tools, high-quality fencing, changes in laws around harvesting, and so forth. But these changes face two challenges. First, they are difficult and expensive. Second, even if they are implemented, animal harm will remain – a route to (all but) eliminate it would be preferable. For both reasons, I propose, it is better to start again. How do we do this? I argue that vertical arable agriculture offers the opportunity to eliminate incidental harm to animals in the harvesting process while still providing us with ample food. It offers us the chance to finally provide a full response to the burger vegan.

While this chapter will be structured around responding to the burger vegan's challenge, it should be clear that these animal thieves are of particular interest in the present enquiry. The relationship we have with these animals is grounded almost entirely in food. These animals visit agricultural land because of a desire to eat those crops that we grow for our own purposes (or to eat those animals who *themselves* wish to eat crops). These crops, of course, are being grown so that they can be fed to us or to other animals[2] – hence my provocative framing of the animal neighbours in question as *thieves*. The pertinent question about these animal thieves is thus not what we may feed them, or whether we are obliged to feed them, as it is with our suburban animal neighbours. It is, instead, a question of how we may *prevent* them from feeding, and – crucially – how we may conduct our agricultural practices in a way minimally respectful to them.

It may seem that I have already provided the normative tools to explore this case. Drawing upon the previous chapter, we may think of these animal thieves as just a particular kind of animal foe.

Like animal foes, then, we can repel them, and we are under no obligation to welcome them into our spaces. This, I think, is right. But it is insufficient as a response to the burger vegan's challenge, which does not claim that we wrong these animals by keeping them out of our fields, but that we wrong them by killing them in massive numbers to produce food. First, we kill them deliberately as a means of 'pest control'. On the face of it, that is easy enough to condemn on animal rights grounds, and will not receive much attention here. (The complicating factor, which must be acknowledged, is that we can raise questions about the extent to which we could effectively feed ourselves – and other animals – without the use of lethal control measures.) Second, we kill them in very large numbers inadvertently, in the harvesting process and through other activities linked inextricably with arable agriculture as currently practised. Indeed, given that these deaths occur because of *arable* agriculture, the burger vegan – fairly! – accuses vegans of being complicit. This means that the vision of a vegan diet as removing one from complicity in animal death and suffering is overly simple. It also means, in the eyes of their critics, that vegans are guilty of a degree of hypocrisy. For this reason, our relationship with these animals has become a pervasive and troubling challenge to veganism.

Burger Veganism

Serious harm is inflicted upon animals in the process of arable agriculture. As noted, some of this can be relatively easily condemned by animal rights theorists. For example, where poison or traps are used to kill rodents eating grain, humane forms of deterrence or catch-and-release traps might be used instead. This may be inconvenient, or expensive, but these costs are what we are to expect if we are taking animals' rights seriously. These concerns do not undermine the animal rights case for veganism, but they do underline its need to be complemented with wider restrictions on violence against animals: even a fully vegan society does not respect animals' rights if animals are still being killed elsewhere in food production.

Some other harms inflicted upon animals, however, are less avoidable. The case in point is the deaths of animals in the field, during the harvesting process itself. As will be familiar to anyone who has been present during a harvest, farm machinery kills, maims, displaces, exposes, and – in these and other ways – *harms* the animals who live on, in, and around farmland. Other agricultural practices (tilling, bailing, etc.) also involve harm to animals. But let us focus on harvesting for the purpose of illustration.

The burger vegan's challenge is the following. There are many ways we could gather food for human (and nonhuman) consumption. *All* of these involve harm to rights-bearing animals. Given that we need food,[3] we must select one of these methods. But it is *not* clear that a conclusion about the merits of vegan agriculture is the one that follows from animal rights principles. Steven Davis (2003) compares data about numbers of animals killed and the nutritional value of food produced, and argues that beef from pasture-raised cattle may involve less harm to rights-bearing animals than certain forms of arable agriculture. Tom Regan – Davis's representative animal rights theorist – is committed (Davis says[4]) to taking a path of least harm when all paths available are harmful. Thus, Davis argues, Regan (and other advocates of animal rights) should be prepared to support a mixed beef-and-plant-based food system rather than a purely plant-based food system. Davis's defence of animal agriculture takes on a different form than those prevalent in previous decades (Lamey 2007) – specifically, Davis defends meat-eating *even despite* an acceptance[5] of the normative principles underlying animal rights.

It is worth stressing what the burger vegan's claim about the 'path of least harm' means. It does *not* mean that animal rights reduces to crude utilitarianism. It just means, as deployed by Davis, that we should commit to taking the path that causes the least harm if harm is unavoidable. If the choice is between eating the products of animal agriculture or the products of arable agriculture – given that both involve harm – the rights theorist should favour the route that produces the least harm.

Davis and others supporting the consumption of pasture-raised ruminants (e.g., Archer 2011b; Schedler 2005) offer just one strand of burger veganism. Burger vegans might also talk about hunting (Cahoone 2009; Archer 2011b). But a slightly different trajectory of burger veganism supports the human consumption of those kinds of non-vegan, but (comparatively) harm-light, products pointed towards in chapter 2. Here, the products defended may *already* be compatible with animal rights. The burger vegan arguments are there to bolster the case for these products, rather than as apologism for something *prima facie* incompatible with animal rights (see Abbate 2019b; Fischer 2018; Milburn 2016a). So, for example, burger vegan arguments have been deployed in defence of (human) consumption of the likes of roadkill (Bruckner 2015), invertebrates (Fischer 2016a), and the products of cellular agriculture (Milburn 2018). The thought is that not only are these products (potentially or theoretically) relatively harmless, but the products of arable agriculture are *not* relatively harmless. The case for their consumption is thus bolstered. While arable agriculture continues to inflict harms upon animals, one will potentially be able to construct a burger vegan defence of the consumption of any animal product that can be acquired with no harm to animals, and perhaps even many that can be acquired with only comparatively small harm to animals (see Fischer 2019b).

A kernel of the burger vegan critique has captured the imagination of a non-academic audience. One burger vegan authored a widely shared article called 'Ordering the Vegetarian Meal? There's More Animal Blood on Your Hands' (Archer 2011a). This has led to a range of popular critiques of veganism that, whatever their overall philosophical respectability, have at their centre the powerful burger vegan challenge, with commentators utilizing burger vegan arguments to justify their own non-veganism, their own hunting, or their own husbandry (see, e.g., Woginrich 2014; cf. Francione 2014). Offering a vegan response to burger veganism is thus important for reasons other than mere academic interest.

Three Responses to the Burger Vegan

There are, I think, broadly four responses open to animal ethicists who wish to rebuke burger veganism. Three have predominated in the literature. It is, however, the fourth that I suggest has the most potential. Before getting to this fourth point, however, it is worth exploring the first three, revealing their strengths, but also – crucially – their weaknesses. One can: (1) Bite the burger vegan bullet; (2) Dispute the burger vegan numbers; or (3) Challenge the philosophical assumptions of the burger vegan.

Biting Bullets (and Burgers)

One could accept that the burger vegan is right that on an animal rights view, we should be eating (some) grass-fed beef rather than relying (entirely) on industrially harvested crops. While it may sound like this means the vegan accepting defeat, this position is well suited to responding to popular articulations of burger vegan logic. This is because – despite pithy headlines – burger vegan conclusions offer little support for a typical Western high-meat diet over a typical Western vegan diet. If anything, they offer further arguments in favour of veganism, as intensively raised animals are themselves fed the products of arable agriculture. Given that it is more efficient (i.e., less harvesting is required) to feed (say) soy directly to humans than it is to feed it to animals and then later kill the animals for meat, burger vegan arguments indicate that a typical Western high-meat diet is even *more* destructive of animal life (compared to a typical Western vegan diet) than it initially appears. Not only are non-vegans responsible for the deaths of the farmed animals whose bodies they eat, but they are responsible for the deaths of the animals killed on the arable farms producing food for those farmed animals.

As such, the burger vegan 'proposal is radically unsupportive of the *status quo* in animal agriculture. In effect, [the burger vegan] proposal calls for the complete abolition of intensive confinement and an end to poultry and pork production' (Matheny 2003, 506n1). Perhaps the ideal suggested by the burger vegan argument is an almost-vegan who consumes a little beef from pasture-raised cattle and animals painlessly hunted. It should be clear that the vegan is far closer to this ideal than is the typical Western meat-eater – or even 'conscientious' non-vegans favouring 'humane' meat.[6]

Perhaps the vegan could concede that she has only a slight quarrel with the burger vegan – the burger vegan is certainly not the friend that the unrepentant meat-eater might assume. Biting the bullet not only disarms the interlocutor, but turns her own argument against her, insofar as she now finds herself having to justify her consumption of animal products *other* than grass-fed beef (or meat from humanely and sustainably hunted kangaroos [Archer 2011b], or similar). And if the interlocutor does *not* engage in these other forms of consumption, then, again, it is not clear that she has any great quarrel with the vegan in the first place. Both share a common opponent in the paradigmatic Western meat-eater.

One may think that the trouble with biting Davis's bullet here is that it blunts the edge of veganism by allowing the consumption of non-vegan products, and concedes that one can never be 'fully' vegan. I do not think that this is a problem; indeed, I hold that animal ethicists, even while advocating veganism, should allow that there are certain edge cases in which the consumption of animal products is not ethically problematic – or, at the very least, not *grossly* ethically problematic (Milburn 2016a; cf. Abbate 2019b; Fischer 2019b) – and should allow that, in the world as it exists, it is next to impossible to be a 'perfect' vegan (Gruen and Jones 2015). This point will be returned to later.

Thus, we must distinguish between different burger vegan bullets, and note that some are far less appealing than others. Even if some animal advocates are willing to bite bullets (if bullets they are) concerning roadkill, insects, or cellular agriculture, it seems far less likely that they should be ready to accept cattle farming, hunting, or ranching. Considered from the point of view of the cow (or whoever), an 'animal rights' position allowing cattle farming, hunting, or ranching seems to have been watered down to the point that it is not one worth supporting. At the very least, we have a responsibility to explore other alternatives.

Thankfully, however, there are good reasons for vegans to resist the biting of these less appetizing bullets, as will be explored over the following sections.

Disputing the Burger Vegan's Numbers

Despite the influence and significance of Davis's argument, there is a fatal flaw in his analysis of the data. As an early critic pointed out, Davis mistakenly assumes that 'the two systems – crops only and crops with ruminant-pasture – using the same total amount of land, would feed identical numbers of people ... In fact, crop and ruminant systems produce different amounts of food per hectare' (Matheny 2003, 506). Once Davis's analysis is modified to account for this, 'we see that a vegan-vegetarian population would kill the fewest number of wild animals, followed closely by a lacto-vegetarian population' (ibid., 507) – neither of which is Davis's preferred population.

Further, Davis's claims about the number of animals killed are, at best, highly provisional. For example, one study Davis uses followed a mere thirty-three mice, of whom *one* was killed during a wheat harvest. Eight mice were then followed during bailing, and none were killed. Two out of five mice followed were killed during stubble-burning – which is far from a universal practice. Davis, however, uses a figure of 52 per cent of mice being killed. Of the thirty-two mice remaining after the single death in the first study, seventeen were (assumed to be) killed by predators. So, one death from harvesting, plus seventeen deaths from predators, gives a death rate of eighteen out of thirty-three, or 52 per cent (Lamey 2007, 335–6). However, the inclusion of the seventeen predated mice in these numbers is deeply controversial. Regan (who, recall, is Davis's example animal rights theorist) argues that wild animals cannot violate rights, so it is far from obvious that a predated mouse has her rights violated. They are certainly not violated by the predator. Might they be violated by the farmer? While the farmer may have a degree of causal responsibility for the deaths, as argued in the previous chapter, the farmer would *not* owe any particular duty of protection to the mice, unless (implausibly) she had encouraged their presence for some reason. Consequently, it is a stretch to say that the rights of these mice have been violated. Unless wild animals possess a general right to assistance in the face of predation – which is a bold claim (see chapter 7) – it is hard to see how the rights of the mice have been violated.

Further serious problems with the numbers used by Davis and Archer are outlined at length by Bob Fischer and Andy Lamey (2018). For example, Davis fails to account for the fact that the sugarcane farming he explores (as an example from which to extrapolate the number of field-animal deaths) takes place over a biennial cycle, thus (presumably) he should halve the number of deaths that he predicts based on an annual cycle. Archer, meanwhile, clumsily conflates the likelihood of mouse-population explosion over a large area (such as the Australian state of Victoria) with the likelihood of a mouse-population explosion on a single arable farm.

This critique can take us only so far. These mistakes could be corrected, and new figures could be calculated. The trouble is that we simply do not have the data available to come anywhere near accurately identifying death rates. Fischer and Lamey (2018, 416) identify a total of three peer-reviewed studies on the topic, which provide for a total of thirty-one deaths. Fourteen of these come from the just-mentioned sugarcane study, which documents deaths resulting from an agricultural method now extinct. And there is no reason to think that this tiny dataset offers conclusions that are generalizable, given the many animal species, habitats, crops, agricultural techniques, and so on that must be accounted for. Even if existing burger vegan arguments rest on limited data that has been badly interpreted, we cannot assume that future data will not be more friendly to the burger vegan case.

There is a further sense in which this response to the burger vegan can take us only so far. Even if Fischer, Lamey, and the rest are completely correct, while Davis, Archer, and the rest are completely mistaken, the problem is not settled. All that the burger vegan need do is identify some other source of animal foodstuffs that entails less harm, and/or some other source of arable agriculture that entails more harm. For instance, whatever the merits of Davis's own calculations, it seems hard to argue with Donald Bruckner's conclusion about roadkill if we are playing a straightforward numbers game. Surely, Bruckner could say, I will be responsible for fewer animal deaths were I to pull over and pick up a recently struck doe than buy the equivalent weight (a doe will yield, I am led to believe, between ten and fifteen kilograms of edible meat) of tofu, given the expected impact of farming soybeans. If one seeks to respond in a compelling way to burger veganism generally, one needs to do much more than dispute the numbers of its various proponents.

What is indisputable – whatever the details of published burger vegan cases – is that contemporary arable agriculture causes animal deaths. Mammals, birds, amphibians, reptiles, fish (from fertilizer/pesticide run-off) and – though this raises too many ethical puzzles to address properly in the present context – a colossal number of invertebrates are killed by arable agriculture. While a combination of bullet-biting and number-disputing can serve to rebut many extant burger vegan challenges, they cannot change the fact that animals are being killed by our farming practices.

Challenging the Philosophy of Burger Vegans

The philosophy underlying the burger vegan's proposal may leave much to be desired. Take Davis as an example. Despite an initial concern with *harm* to animals, he quickly switches to considering solely *animal deaths*, and it is hard to imagine an ethical system that is concerned solely with deaths and not other kinds of harm (Matheny 2003, 507–8). Note that the system with greater death could involve less harm, and so should thus be favoured. (Which involves more harm: torturing twenty dogs for a decade, or painlessly killing twenty-one dogs? The latter contains more death. The former contains more harm.) Thus, comparing the amount of death in a plant-based or plant-and-grass-fed-cattle-based system is only *one* step in comparing the harm inflicted by the two systems.

If we were to move beyond simply counting the number of deaths, there is every chance that we would find that a vegan food system would be significantly less harmful than Davis's preferred system. Let us, like Davis, turn to Regan's rights view. Regan (in)famously contends that a dog has less to lose in death – a dog is *harmed less* by death – than a paradigmatic human, and so he would be willing to throw a very large number of paradigmatic dogs out of a hypothetical sinking lifeboat before throwing out a single paradigmatic human (Regan 2004, 324–7). Similarly, perhaps a paradigmatic cow has more to lose in death than a paradigmatic mouse, meaning that Davis (who follows Regan) should be prepared to see a very large number of mice crushed under a harvester's blade before he sees one cow strung up in a slaughterhouse.[7] Of course, that Davis does not take account of these worries does not mean that he *could* not, and does not mean that other burger vegans *do* not. And the internal problems of Davis's approach do not preclude the success of those utilizing burger vegan approaches in non-rights-based frameworks.

We can also raise questions about Davis's reliance on the idea of least harm. For the rights theorist, there is a big difference between rights-violating harm and non-rights-violating harm. I am harmed (my interests are set back) if I lose a large amount of money in a game of cards. That does not mean that an onlooker should help me cheat, even if my winning this hand will cause less harm to the loser than my losing it would cause to me. This is because the loser is wronged if I win by cheating, but I am not wronged if my opponent wins fairly. If reducing harm is all that matters (as it is for certain kinds of consequentialists), then cheating would be the right thing to do. But it is not all that matters for most of us, including rights theorists. The burger vegan could respond that they are not crudely conflating all forms of harm, as the word *harm* (as they use it) refers specifically to the *wrongful* setting back of interests. But then we could accuse the burger vegan of begging the question by assuming that the incidental killing of field animals entails the *wrongful* setting back of interests. Slaughter wrongs a cow, but it is, perhaps, an open question whether animals are wronged when they are killed accidentally.[8]

This leads us to probably the most significant philosophical challenge to the burger vegan. The *doctrine of double effect* entails, roughly, that it is worse to cause intentional harm (for example, by slaughtering cattle) than to cause unintended-but-foreseen harm (for example, by running over mice). This response has the potential to completely undermine the burger vegan's challenge (Lamey 2019, chap. 4). It is important to note that while the doctrine of double effect may allow the vegan to argue that nonintentional harm to mice is preferable to intentional harm to cows, it need not deny that the harm to the mice is morally significant, nor even that non-intentional killing of mice isn't rights violating.

For the purposes of this chapter, I will remain neutral on whether this vegan response to the burger vegan is successful. My reasons for this are twofold. First, the doctrine of double effect is controversial. Second, there is a sense, I believe, that the response misses the point. Even if *these* harms towards animals should be preferred to *those* harms to animals – and even if *these* harms do not violate rights – we might have very good reason to minimize them. We remain causally responsible for them, and overlooking them (even though I do not accuse advocates of the doctrine of double effect of doing this!) might display a certain callousness. Rather than deploy philosophical theory to justify (even if regrettably) the causing of harm, I suggest that there is considerable value in exploring practical steps we could take to remove the harms altogether.

A New Response to the Burger Vegan

We can see how an appeal to the doctrine of double effect is not, alone, a sufficient response to the burger vegan's challenge if we shift perspective from considering what we should be eating/purchasing to thinking about the issue as a matter of collective decision-making, and especially if we think about this from the point of view of an animal thief.

A deeply philosophical field mouse might be prepared to accept that, if push came to shove, humans might owe a lot more to cattle than to her – that cattle have more to live for than mice (this supermouse presumably excepted), that humans have an obligation to feed those cattle they brought into the world more so than they have to feed her, that humans have a right to harvest crops for their own consumption, and so forth. She might also be able to accept that the doctrine of double effect means that, if there is a choice between deliberately killing a cow and incidentally killing a mouse, humans should choose to incidentally kill a mouse (all else equal). But all this, she might insist, is by the by. Surely, we should want to reach a situation in which we do not have to make these hard choices – in which we do not have to kill cattle *or* mice.

Cochrane, in his discussion of the burger vegan's challenge, begins to point us in this direction. 'Political communities', he writes, 'have an obligation to take greater steps to reduce the number of animals killed in the field' (2012, 101). That this is an obligation of *communities* should be clear; there is little that individual consumers will be able to achieve in this area, and expecting farmers to shoulder the costs of these efforts would be unjust. Cochrane's recommendations in this area are, however, underdeveloped. He calls for 'more efficient land use, more efficient crop strains, more sophisticated fencing, more sophisticated scare tactics, and better management of overproduction', which 'would all serve to reduce the number of animal deaths caused by our agricultural practices' (ibid.). He also points (101–2) to the need to include questions about human population growth in questions about animal death in arable agriculture. While this is not something I wish to discuss at any length, it is true that lowering the human population (or, indeed, the population of animals needing to be fed) could help to lower the deaths of animals in arable agriculture, and that harm to animals in arable agriculture could provide a case for limiting human (and animal) population growth.

Let us look to the practical suggestions about agricultural policy and practice that Cochrane offers. Improving land-use efficiency and developing crop strains that are more efficient are things that farmers (and society) already have a strong incentive to do. Increased land-use efficiency will save farmers money, help feed the world's growing population, and free up land for other purposes. If it is the case that animals live in the same densities in land-efficient farms and land-inefficient farms – an empirical question, the answer to which cannot be taken for granted – then land-efficient farming will see fewer animals put at risk of harm during harvesting and other agricultural activities. Similar may be true when it comes to efficient crop strains – depending, precisely, on what is 'efficient' about them.[9] Agricultural policy as formulated frequently encourages overproduction, by, for example, offering farmers a guaranteed price for their produce, or else having governments buy out excess stock to keep farmers afloat. Again, we already have reasons to be worried about these kinds of policies, but the fact that they are contributing to the unnecessary deaths of field animals provides further concerns.

Cochrane's other suggestions – more sophisticated fencing and scare tactics so that field animals are kept out of agricultural spaces in the first place and/or can be removed prior to harvest – also offer promise, though the precise details are going to vary considerably depending upon the crop, the animals, and the landscape. On the normative framework advanced here, it would be legitimate for us to engage fencing, scare tactics, and other methods to keep field animals out of spaces that humans utilize for farming purposes. These animals are our neighbours, and, as argued in the previous chapter, we have no moral duty to extend hospitality to *particular* neighbours. We are within our rights to take measures to ensure that neighbours to whom we do not wish to extend hospitality are excluded. Indeed, in this case, that is the *responsible* thing to do, given that these animals face extreme risks at our hands if we fail to do this. A failure to extend hospitality in this case becomes morally optimal. Indeed, it becomes morally required, because (as argued in the previous chapter) if we choose to extend hospitality, then we have an obligation to provide a safe environment. And a field that will soon be upturned by a combine harvester is not a safe environment.

A little insight and care can go a long way in preventing harm to field animals. For example, the Royal Society for the Protection of Birds has had considerable success in saving the lives of corncrakes[10] by advising different harvesting techniques:

> Corncrakes prefer to stay hidden among tall vegetation and are reluctant to flee into the open. It was found [during the twentieth century] that up to 60% of flightless corncrake chicks were being lost each year, when they became trapped in the centre of fields by mowers that were working from the outside edges in … However, by creating areas of cover along the outer edges of fields and reversing the pattern of mowing to work from the inside out, corncrake chicks could be flushed to safety during mowing rather than being killed. In 1992 a programme of payments to farmers to adopt 'corncrake friendly mowing' and to delay mowing until August, when the breeding season has passed and the chicks are older, was first implemented in Scotland by the RSPB. This has since been translated into government-run schemes … After these conservation measures were introduced, corncrake populations bounced back thanks to the support of farmers and crofters. Our last Scottish survey in 2016 put their numbers at 1059 calling males – an increase of 165% since the 1990s. (Barrett 2017)

Though this is about the conservation of a charismatic and iconic bird rather than reducing death and suffering,[11] it demonstrates how openness and ingenuity can save animal lives, as well as how the costs of protecting animals in the field need not fall entirely on the shoulders of farmers. Fischer and Lamey (2018, 424–5) address similar ideas. Though acknowledging that the impact of tilling on animal neighbours is contested, they point to the growth of no-till agriculture. Meanwhile, the success of conservationists in seeing certain forms of pesticides banned provides them with hope.

Measures of this sort have genuine potential for alleviating the death and suffering of certain (kinds of) animals on particular (kinds of) farms. With a little innovation from researchers, openness on the part of farmers, and willingness to offer support on the part of the wider community, arable agriculture could gradually work towards alternative models in which the death and suffering of field-dwelling animals is minimized. We can grow food for our purposes, and prevent this food from being eaten by rural animal neighbours, in a way respectful to the rural animals who do (or would) live in and around our agricultural spaces. Not only does this take the wind out of the sails of the burger vegan challenge to veganism, but the positive achievement is threefold: (1) An ethical issue that arises when it comes to feeding ourselves has been resolved; (2) An ethical issue that arises when it comes to feeding animal members of our community has been resolved; and (3) An ethical issue that arises when it comes to appropriately relating to a particular group of animal neighbours has been resolved.

The trouble is, though, that alleviating death and suffering is not the same as eliminating death and suffering. It is, I fear, a stretch to imagine that developments like this will make food production harm-free – or even close to it.

A More Radical Alternative

Animal ethicists who take minimizing incidental harms to animals in the field seriously see them as 'fixable problems: we just need better regulation and further efforts on the part of animal advocacy groups' (Fischer and Lamey 2018, 416). My worry is that if these regulations and efforts look like those discussed in the previous section, then the problems are fixable only in the sense that we can *reduce* the imposition of death and suffering. Carefully crafted legislation, concerted efforts by animal advocacy organizations, and more research could lead to a multitude of changes to arable agriculture like those that have protected the corncrake. This would not be for the sake of preserving a charismatic species, but for the sake of protecting rights-bearing individuals.

Such advances would require changes to the kinds of topics considered worthy of political discussion. Given that deaths in arable agriculture (beyond an occasional response to burger vegan critiques of veganism) are basically off the table for even animal activist organizations, such a change may seem difficult to achieve – and not cheap. If we were to implement this change as a society, and provide support to arable farmers to make a whole host of changes to make their farming more animal-friendly, we would be able to make arable agriculture significantly more humane. And this would be something to be celebrated. But I contend that these kinds of measures would not, alone, go far enough. While these kinds of tweaks to existing practice may have real merit as means of making our world today more just, I believe that our long-term goal should be rethinking arable agriculture altogether.

Let me put this another way. Animals live in agricultural land. This is practically a truism – not least because there is, basically by definition, food there. And animals are going to be killed and suffer when we drive harvesters over the land on (or in) which they live and feed. Better fencing, sophisticated technological descendants of scarecrows, and changed harvesting practices will get us only so far. Would it not be better to rethink farming from (with no pun intended) the ground up? It is vertical agriculture that gives us this opportunity.

The central, simple idea of vertical agriculture is that farming is taken from two dimensions to three dimensions. More space is created not by expanding farms outwards, but by expanding them upwards. As such, a wide array of agricultural practices could meaningfully be described as forms of vertical agriculture (Beacham, Vickers, and Monaghan 2019). Many of these are already deployed in various places. Some are theoretical. Some utilize new technologies. Some do not.

Certain forms of vertical agriculture can be put aside for present purposes. Rooftop gardens, 'living walls' (i.e., crops grown out of or on buildings' walls), and balcony gardens can all be described as vertical agriculture. These need not utilize any particularly unusual technology, and are (in principle) accessible to anyone with access to a rooftop, balcony, or wall in a not-too-unforgiving (micro)climate. To be sure, this sort of small-scale, unmechanized food production is unlikely to involve much harm to animals, just as the production of home-grown, organic, low-tech crops is unlikely to involve much harm to animals.[12] It might still involve *some* harm to animals – even when we discount the use of poisons, traps, and such. The vegetable gardener in her allotment or garden is always at *some* risk of crushing or smothering ground animals when she pours compost onto a bed, always at *some* risk of trapping birds when she puts up netting, and always at *some* risk of drowning burrowing animals when she drenches her vegetables. The equivalent is true of the urban gardener on her balcony – though the risk is, admittedly, low. That said, it is unlikely that these forms of vertical agriculture have the capacity to make much of a dent in, let alone replace, contemporary industrial farming, not least because they are more suited to growing for one's own consumption, and options are often very limited. Some lettuce in summer is one thing, but providing a variety of fruit and vegetables all year round is possible only with a good deal of space and expertise. A grow-your-own ethic – even while it might be commendable as a way to supplement one's own diet (and to, however slightly, reduce one's reliance on industrial agriculture) – is not really what I mean by shifting towards vertical agriculture. Nor, indeed, is it what is typically meant in discussions of vertical agriculture.

Vertical farming in its more radical sense refers to farming over multiple levels in a controlled environment (Beacham, Vickers, and Monaghan 2019). It might have little resemblance to contemporary arable agriculture (which, we must remember, might have little resemblance to people's *perception* of contemporary arable agriculture). Vertical farming might do away with soil in favour of more efficient methods of feeding plants. Many vertical farms are *hydroponic*, growing plants directly in water. This is not new technology, and replicates or builds upon food-production techniques that have been in use for decades. Drip-based hydroponic systems see tubes dripping nutrient-rich water directly at plants' roots; raft-based hydroponic systems see plants grown on an artificial substrate floating on nutrient-rich water; or plants can grow with a thin layer of solution around the roots. In contrast, *aeroponic* systems see roots hanging in the air and misted with a small amount of nutrient-rich water. Each of these techniques allows layers of crops to be grown one on top of the other. (These are just the start; a range of other soil-free or soil-light systems are in use in vertical or otherwise indoor agriculture.)

Vertical agriculture utilizing these various growing techniques can be employed in many kinds of structures. Traditional (or not-so-traditional) glasshouses allow some natural light.[13] Given the vertical layering of plants, however – meaning some layers are in the shadow of others – glasshouses used in vertical agriculture need to employ either some system of rotation, some use of artificial lighting, or both. Other kinds of structures (including structures below ground) offer little or no natural light for crops, but they can nonetheless be turned into 'Plant Factories' using artificial lights. (See below for a discussion of energy use.)

The greenhouse, former warehouse, or disused train tunnel stacked high with shelves of lettuce and microgreens is a typical example of contemporary vertical agriculture. However, it is not the most evocative image raised by the prospect of expanding growing space skywards. This, instead, is the agricultural skyscraper: a multi-storey building, perhaps located in a city centre, in which most or all floors are given over to growing crops. At present, the vision is hypothetical. But such a farm could be a *highly* efficient use of land, in terms of agricultural productivity. Though these numbers are theoretical, Dickson Despommier (the academic champion of vertical agriculture) and his students have offered rough numbers for the food a single farm could produce (Despommier 2009, 86). Take a thirty-storey building occupying one Manhattan block – so about five acres at ground level. On the face of it, such a building would provide 150 acres of agricultural land (five acres multiplied by thirty storeys). This is very little if one is seeking to feed a city. In fact, however, the productivity of each five-storey floor would be higher than traditional farmland. The land is productive all year; fast crops like lettuce can be ready in weeks (at any time of the year, not just the height of summer), while slower crops traditionally only offering an annual harvest can have multiple croppings. Crops can also be grown more densely, and multiple layers of crops can be grown per floor. Factoring in these various advantages, Despommier contends, multiplies the 150 acre figure by a factor of sixteen; quadruple the croppings, double the density, and two layers per floor. This means that the tower could produce the same as 2,400 acres – around three times the size of Central Park. Moreover, even this number could be conservative. It does not account for the possibility of twenty-four-hour lighting; it does not account for the loss from spoilage in transport from rural farms to urban centres; it does not account for losses on rural farms attributable to disease, flooding, drought, and so forth; and it does not focus only on those crops (currently) best suited to vertical agriculture.

The key advantages of vertical agriculture, according to the academics working on the question (see, e.g., Benke and Tomkins 2017; Kalantari et al. 2018), relate primarily to the more efficient use of land and other resources, and the greater food security offered. There are a host of other possible advantages (social, economic, environmental, etc.) identified, but, in these discussions, animals' interests are noticeably absent. Perhaps the closest that vertical agriculture's advocates come to mentioning benefits to animals is an environmentalist concern with rewilding farmland. The idea is that because vertical agriculture is a considerably more efficient use of land for food production than (even intensive) 'traditional' forms of arable agriculture, large swathes of land formerly used for farming can instead be given over to native ecosystems, including native animals. Now, of course, such rewilding would not happen automatically (Tuomisto 2019, 276), and so the proposal to switch over to vertical farming is often combined with a proposal for financial incentives for existing owners of agricultural land to rewild. Indeed, such a proposal also serves as a response to worries about the impact on rural economies and the livelihoods of farmers – in principle, farmers could be better off after a widespread shift to vertical agriculture if they were offered new (fulfilling) work caring for the land (Platt 2007, 84).

However, the benefit that leads for me to call for vertical agriculture – the removal of incidental harms to animals caught up in agricultural processes – is completely absent from existing discussions of vertical agriculture.

Interestingly, though harms to field animals are absent from existing discussions, the animals themselves are not. Advocates of vertical agricultural technologies note major advantages associated with excluding agricultural 'pests'. The main pests referred to are diseases of various sorts, reflecting the main threats to crop security faced by those in the affluent West. They also, however, mention the *animal* 'pests' – thieves – who can be easily excluded from closed, indoor agriculture (Despommier 2013, 388). Vertical agriculture is to be sealed off from the outside world. Indeed, hermetic, sterile environments are already utilized in some commercial forms of arable agriculture. As such, there is little chance for bacterial, viral, and fungal pathogens to enter these farms – let alone mammals, birds, reptiles, amphibians, and the others who are crushed by combine harvesters. Thus, no animals are at risk from harvesting processes. Vegans consuming the products of vertical farms (or feeding them to other animals) could, contrary to Archer's provocative headline, genuinely have no blood on their hands.

My proposal, then, is this: if we are serious about animal rights, we should, in the medium to long term, support a shift away from conventional forms of arable agriculture towards vertical agriculture. This is because vertical agriculture can do away with the incidental harms inflicted on animals in the harvesting process.

Note that this is very much a *social* or *political* proposal. It would be unreasonable and impractical to ask individual consumers to bear the burden of a shift to vertical agriculture. Most individual consumers can do little to change their consumption and purchase habits to favour vertical agriculture rather than conventional forms of arable agriculture. They have some reasons to reduce their reliance on the harm-causing industry of arable agriculture (cf. Bobier 2020), but options are limited. Very few could turn entirely to vertical agriculture. Some consumers could favour home-grown food, purchase the products of vertical/indoor agriculture when they are available, and take steps to limit consumption (e.g., reducing food waste). Their primary responsibility in relation to field animals would be pressing for institutional change. At this early stage, it may involve simply raising awareness of vertical agriculture or encouraging animal-activist organizations to get behind it. In the future, such a duty may manifest in putting pressure on politicians to favour vertical agriculture, offering support (donation and investment, for example) to groups developing vertical agricultural technologies, and, of course, purchasing the products of vertical agriculture rather than the products of (currently) typical modes of industrial arable agriculture.

Challenges to Vertical Agriculture

To conclude, it is worth responding to some possible challenges to my call for a shift towards vertical agriculture.

Vertical Agriculture Is Impractical

Much of the existing ethical discussion around vertical farming concerns whether vertical agriculture could genuinely feed the world. As my focus here is comparatively narrow, there is not space to address all these worries. However, it is worth indicating what some of these worries are, and how they might be overcome.

First, high start-up costs could be offset by government support, by favouring the regeneration of derelict sites (Benke and Tomkins 2017, 20–1), and by moving the vertical farm out of the city (Beacham, Vickers, and Monaghan 2019, 280) – there is no inherent reason that vertical farming needs to take place in urban centres. Indeed, it can take place anywhere where water and energy can be supplied (Despommier 2009, 83). For example, perhaps vertical agriculture could take place on sites unsuited to 'traditional' forms of animal agriculture, undercutting challenges from meat apologists that certain land is 'only' suitable for pastoral agriculture. (It is worth adding that there are many derelict sites outside of cities; plenty of provincial towns need regeneration.) It is true that moving vertical agriculture out of the city could slightly increase its environmental impact due to an increase in the need for transport, but only a small part of the environmental impact of food production comes from transport anyway; the loss would be minimal.

Second, energy costs (both financial and environmental) can be offset by the development of efficient technology and through incorporating energy production in farm design. Many designs, for example, include the capture of solar and wind energy, the recycling of grey or black water, and ultra-efficient LEDs. In a 2017 review, Kurt Benke and Bruce Tomkins concluded that 'the external energy requirements of an indoor farm have diminished greatly and are likely to approach off-the-grid operation at some time in the near future' (2017, 21). So while (for example) the lighting requirements of indoor agriculture sound deeply energy intensive, innovative design could, in time, more than offset these costs.

Third, though a relatively small range of crops are currently favoured by vertical agriculturalists – such as salad crops and strawberries – this is because of economic decisions, not because other crops are unsuited to vertical agriculture (Benke and Tomkins 2017, 21). Were the economic incentives to change, the farming of these other crops in vertical arable agriculture would become desirable. For example, were economic incentives provided to support the development of vertical agriculture – particularly vertical farms growing those crops that currently lead to the most incidental harm to animals – no doubt the private sector would rush to provide methods and cultivars suited to growing (say) fruit trees, root vegetables, and commodity crops. (This is not to say that there are not *already* methods for these things. There are any number of soil-light systems demonstrably able to effectively grow potatoes, for example, which are sometimes assumed to be unsuited to vertical agriculture.) There is, ultimately, no limit to what can be grown vertically (Despommier 2007, 82).

Vertical Agriculture Is Not Solely Arable

Some less-discussed visions of a future urban vertical farming system include forms of animal agriculture alongside the growing of crops. Discussing his early designs, Despommier writes that the list of potential food sources 'included poultry, fish, crustaceans and mollusks. We didn't go as high up on the food chain as mammals, because some of our students had an aversion to that particular group' (Platt 2007, 83). The 'aversion' cryptically referred to presumably relates to the dystopic vision of pigs[14] not only crammed into cages in a warehouse, but *stacked one on top of another*. Regrettably, such a vision is likely not far from being realized. Relatively closed systems are used to grow the likes of 'minimal disease' pigs (Evans 2019, chap. 3), while a Chinese firm manages multi-storey 'hog hotels' to produce pork (Davis 2018). Combining these technologies would approximate, no doubt, Despommier's vision. He is no animal rights advocate (Platt 2007, 87). It should go without saying that, from an animal rights perspective, the use of vertical or controlled-environment farming techniques for animal agriculture must be condemned – at least, when the animals in question are rights-bearers.

Another mention of animals in the literature on vertical agriculture is a little more ambiguous from an animal rights perspective. Some forms of 'ponic' farming make use of animals as a part of the process of growing plants. Aquaponic systems combine aquaculture and hydroponics, raising fish for slaughter and using their faeces to fertilize plants. Naturally, an animal rights perspective will condemn the slaughter of the fish, but whether a slaughter-free aquaponics system could be practical, desirable, and rights-respecting is an interesting question worth exploring – though not one that we presently have space to address. Vermiponic systems, meanwhile, make use of vermiculture, which is composting using worms.[15] This, too, raises interesting ethical questions – especially as there is an open question whether worms are rights-bearers. Aquaponic and vermiponic systems are far from the norm in existing and proposed vertical agricultural systems, however. For the purposes of the present argument, it can be assumed that I am talking about more standard hydroponic, areoponic, or other (non-animal-based) systems. However, I here leave it as an open question whether forms of aquaponic and vermiponic systems are acceptable from an animal rights perspective. I hope such analysis will be forthcoming.

Further, vertical agriculture may implicate animals in less direct ways. Unlike other forms of arable agriculture – aquaponic and vermiponic systems aside – it is unlikely to rely on animals as a source of fertilizer. Nonetheless, any number of inputs may be animal-based, or may have involved animal testing. There is a good chance that the development of the nutrient-rich waters of aquaponics and aeroponics will have involved animal testing at some point, and all kinds of building materials may contain animal products. For instance, both drywall and paint often contain animal products. The reality is that, given the use of animals in our contemporary world, it is hard to imagine a way of living in the contemporary West (short of fully withdrawing from society) that does not involve contact with items containing animal by-products (cf. Gruen and Jones 2015). One certainly would not be able to travel anywhere; cars, even when not decked out with leather seats, contain animal by-products – and even if they did not, asphalt does.

Thus, this problem of animal by-products in the most innocuous of objects is not in any way a problem unique to vertical farming.

The best solution we have at present, which is reminiscent of the consumer-based solutions to a reliance on harm-causing arable agriculture, is threefold: First, we should do what we can to avoid animal products and favour vegan products; Second, we should aim to consume less in general; Third, we can support commercial and legal pressure on animal agriculture, which will gradually result in animal ingredients being significantly less desirable to producers of all kinds. In any case, the widespread presence of the by-products of animal agriculture in everyday objects should not give us reason to oppose production practices that are more animal-friendly – especially without compelling evidence that vertical arable agriculture would entail vastly greater use of animal by-products than other forms of arable agriculture. This is an empirical question, on which (so far as I know) there is no available data – though it is hard to see why it would.

Vertical Agriculture Involves Incidental Harm, 1

Perhaps the most significant challenge to my proposal is that vertical agriculture still results in incidental animal harm, and so does not work as a viable alternative to dominant forms of arable agriculture for present purposes. There seem to be two ways that this critique could be run. The first is that vertical agriculture itself is harmful. The second is that the shift away from current forms of arable agriculture is harmful. Let us address these in order; the first in this subsection, and the second in the next.

First, it could be observed that building practices will cause incidental harms to animals. Clearing brown- or greenfield sites will kill or otherwise harm many animals present in the space, undercutting any claim that vertical arable agriculture would be harm *free*. And the primary industries responsible for the extraction of building materials will harm animals, too. (Note that this is not simply a claim about animal products present in building materials. It is a claim about how construction necessitates a reliance on harmful industries.) Two things can be noted in response. First, these harms can be significantly offset by taking the earlier suggestion of converting existing buildings for the purposes of vertical agriculture, rather than constructing new buildings. Second, even if resource extraction for buildings (and/or clearing space for buildings) is harmful to animals, this is one-off harm. In contrast, existing methods of harming rely on harvesting, clearing, and other interventions indefinitely into the future. Perhaps, nonetheless, this one-off harm will be greater than it needs to be. While we certainly have reason to take steps to make resource extraction and building more animal-friendly, here is not the place for that discussion.

On the face of it, it seems that the construction of vertical farms will result in significantly less harm to animals relative to the amount of food produced than conventional farming practices – if not in the short term, then definitely in the medium to long term. The reason for this is that once the facility is up and running, there is no reason to believe that incidental harm to animals will be anything but negligible.

Vertical Agriculture Involves Incidental Harm, 2

Let us instead turn to the second part of this challenge, which concerns harm to precisely those animals this proposal is aiming to protect. Imagine two possible food systems: one that looks much like our own (though without animal agriculture), and one that relies solely upon vertical arable agriculture. Which would be the one with greater animal suffering? If the arguments of the previous subsection are correct, the vertical agricultural system would entail less harm to animals. Nonetheless, the advocate of a more conventional system might press, the animals who have been 'saved' by the shift to vertical agriculture may have reason to complain. While they are no longer being run over by combine harvesters, they have lost their home. Indeed, they may find that they are dying out in large numbers: their favoured habitat no longer exists, their favoured food has vanished.

However, this is where the framework developed in the previous chapter really shines a light on the particular obligations we have to these animals. While we are obliged to take steps to not harm these animals, we do not have an obligation to help these animals. (Or, at least, it is not obvious that we do.) If a family of mice thrive by taking food from my compost heap or bin, I am not permitted to kill them. At the very least, I am not permitted to actively and deliberately kill them. But I am under no clear obligation to continue feeding them. I could, I contend, permissibly stop putting food waste in my compost heap or bin, or I could fit a better lid, or I could go on holiday, or I could move house. Were I to do any of these things, it is possible that some of the mice would go hungry, forcing them to take significant risks to seek out new food. But there is a distinction between cutting off a food source and foreseeably-but-unintentionally killing an animal with a combine harvester. The former gives an animal a chance to move on, to find a new way of living; the latter cuts off any chance of living. The animals living on agricultural land, too, could find a new way of living. These animals, though presently somewhat dependent on humans, were not spontaneously generated by agricultural land. Their ancestors – who will not have been all that different from them[16] – thrived somewhere else. And they could, too. It is the difference between catching and releasing bats living in a roof and simply demolishing the house. The catch and release strategy says 'No, you can't live in this space, but you can live elsewhere, or stick around as this space becomes something else, if you like the new space'. The demolition says 'We don't care about you'. In short, the one approach is a way to live and let live, while the latter is not. So even if a change in farming practices will result in the death of some animals, this does not mean that we have killed the animals – *either* in the way we kill animals with traps and guns, *or* the way we kill animals with combine harvesters.

Thus we will usually (more on this shortly) have no obligation to continue farming for the sake of the animals who live on farmland, just as I have no obligation to continue throwing food in a damaged bin just because some mice eat out of it. But perhaps rewilding involves harm to animals, either because it creates a habitat unsuited to the animals who *did* live in the agricultural land, or because it creates a new space in which harm to animals takes place. There are, I think, four groups of animals who could be deemed to be 'harmed' by the creation of this rewilded habitat.

First, there are the field animals who exist at the time of the rewilding. As just explained, we are not responsible for these harms. If we have not fostered any dependency they have on us, then we do not do wrong in leaving these animals to fend for themselves.

Second, there are the descendants of these field animals. These are wild animals, and while they may be born into an environment less suited to them than agricultural land, the fact that they have been born shows that it is not *un*suited to them.[17] I contend – as will be discussed in greater detail in chapter 7 – that we are not typically morally responsible for harms befallen by wild animals in the normal course of their lives. The harms that wild animals inflict upon each other (or otherwise experience) are frequently nothing to do with us, even if they occur in spaces that we have had some hand in creating. We thus have no special moral responsibilities to protect these animals from harm, or aid them in times of need.

Third, there are the animals who *would have been* the descendants of the field animals *if* the farmland was not rewilded. But these animals do not come into existence. Not coming into existence is not a harm – that which does not exist cannot be harmed – and so we have nothing to worry about for their sakes.

Fourth, there are the animals *unrelated* to the field animals who come to live on the rewilded habitats. But these, again, are wild animals. We are generally not responsible for the harms they experience.

As such, it seems that arguments that a move away from current forms of arable agriculture and towards forms of vertical agriculture will harm wild animals fall flat.

But what if we *do* have responsibility for the suffering of field animals or their descendants, and thus we *are* responsible for some of the harm that results from rewilding? In some cases, field animals may have been encouraged (and their dependency nurtured) to a sufficient degree that a shift in land use would wrong them. But even if this is true, it is not a counter to my arguments about vertical agriculture. Instead, it is an argument that we need to be careful about the kinds of 'rewilding' we favour. If we would wrong field animals by ceasing to farm the land, then we could continue to 'farm' it *as a wild space*. That is, we could 'farm' it as means to manage the space, just as any number of contemporary protected areas are managed to create habitats for particular animals. This 'farming' would, of course, be carried out *without* harmful harvesting processes, leaving all or much of the food to the field animals. These animals would be thieves no more, but friends.[18] Whether steps like this are necessary – and which such steps are required – is not something that can be mandated from the armchair. It will require close attention to the particularities of the situation.

Concluding Remarks

Vertical agriculture's advocates have had little to say about minimizing harm to animals, and animal advocacy organizations – who generally limit concern about field animals to responding to the burger vegan critique – have said little about supporting vertical agriculture. It is my contention that advocates of vertical agriculture should address its potential for alleviating harm to animals, while animal advocates and vegans should throw their support behind vertical agriculture to (all but) eliminate – not only limit – harms to field animals resulting from harvest. Not only will such a step radically limit anthropogenic harm to animals, but it will completely undercut the chance of a burger vegan challenge.

By looking towards animal thieves, we have taken yet another step away from domesticated animals, and towards wild animals. We do not live with or near these animal thieves, but we nonetheless – inadvertently – feed them. There is thus a minimal level of 'domestication' – these animals can, like the urban animals of the previous chapter, be called 'liminal'. It is now time to take a further step along this path away from domestication and towards wildness. For the remainder of the book, my attention will be turned to animals who are 'wild' in the stricter sense. In the next chapter, I will address wild animals who (for their own sake) are kept in human captivity, at rehabilitation centres. In the final chapter, I will address free-living wild animals.

6

Animal Refugees

Imagine you were to encounter an injured mouse or pigeon, along with a nest of dependent pigeon squabs or mouse pups. Aware that these animals are almost certainly unable to survive on their own and being a firm, altruistic believer in the value of animals, you opt to take them to a wildlife rehabilitation centre (WRC). This is not a sanctuary of the kind favoured by environmentalists, which is concerned with preserving habitat or protecting members of favoured rare species. Nor is it the kind of farmed animal sanctuary supported by many vegans, animal activists, and animal ethicists. Instead, this is a sanctuary environment dedicated to helping individual non-domesticated animals for their own sakes. This WRC, we can imagine, accepts the mice or pigeons brought to them, perhaps with an acknowledgment that all animals deserve a chance. This is the kind of language frequently employed by these institutions when stories about them accepting members of maligned or common species – like mice and pigeons – circulate in the press. *All* animals matter, they say, so they are ready to accept these needy animals, even though they are 'only' mice or pigeons.

But now imagine that, a week later, you contact the WRC to check up on the animals you found. You would surely be horrified to be told that they have been fed to some other resident of the WRC – even if this other resident is a member of a species that, in the wild, would routinely prey upon these animals, such as a bird of prey, or wildcat, or mustelid. 'But,' a WRC volunteer might press, 'we need to feed our residents, and we aim to feed them in ways that replicate their natural diets. If it wasn't the animals you brought us, it would just be some *other* small bird or rodent that was killed to feed our resident. All animals matter – and that includes predators.' Whether such an explanation would quell the initial horror is something I leave for the reader to decide.

This story is fiction. What is not fiction is that WRCs will feed meat to carnivorous and omnivorous animal residents, and that this meat may be meat from animals who are very like the WRC's other residents. There is a tension at the core of the WRC's mission; on the one hand, it is concerned with helping animals who need it – sometimes employing the rhetoric of all animals deserving a chance. On the other hand, the sanctuary directly supports or engages in harm against animals by feeding the rescued animals meat, milk, or eggs acquired in deeply unjust ways.

Animals living in these kinds of sanctuaries occupy a curious conceptual space somewhere between wild and domesticated animals, but very unlike the 'liminal' animals explored in the previous two chapters. Rather than neighbours – friends, foes, thieves – these animals are a kind of refugee, a being from another place who is being helped in her time of great need.[1] These are animals under direct control of humans, and they benefit from this control; but they are not members of our communities. At least, perhaps they aren't *yet*. And WRCs are frequently going to resist the suggestion that their charges are, or should be, members of our communities. WRCs seek to *rehabilitate* their charges. The ultimate goal, in many cases, is to see the animal released into the wild. When this is not possible, it is presented as a regrettable case – or even a failure.

In this chapter, I explore the ethical challenges that are raised by the diets of the residents of WRCs and the feeding practices that the institutions employ. This exploration will advance as follows.

First, I will set out why investigating this kind of human-animal relationship is important – even if it has seen relatively little consideration among animal ethicists. I will then reflect on the varied meanings of *rehabilitation*, and show that these various meanings bring with them two distinct ethical challenges related to animals' diets.

Second, I move on to consider the first kind of rehabilitation – rehabilitation *qua* restoring health. A crucial part of this, and a central ethical dilemma for sanctuaries, concerns feeding. It will be my claim that wildlife rehabilitation cannot involve the feeding of animal products that have been acquired in an unjust way.

Third, I will consider two other kinds of rehabilitation – rehabilitation *qua* restoring good functioning, and rehabilitation *qua* restoring moral propriety. It is my claim that wildlife rehabilitation centres are making a moral mistake if they seek to re-release animals who will hunt and kill other rights-bearing animals. While wild animals' hunting and killing of other wild animals does not generally – I hold – involve injustice, humans can become responsible for the death and suffering of animals if they facilitate predation. At this time, such hunting *can* involve an injustice. It is my claim that the death and suffering inflicted by released predators leaves metaphorical blood on the hands of rehabilitation staff. Thus, the only appropriate move for sanctuaries to make is to restore the health of predators with the intention of keeping them indefinitely in a sanctuary environment (or similar). The chapter concludes with some reflections on what these conclusions mean, both for theory and practice.

Why Rehabilitation?

Let us put to one side what we might call conservation centres and sanctuaries for domestic animals of various kinds as different sorts of institutions from WRCs. These have different motivations, and raise different ethical issues. We might think that WRCs are of limited theoretical or practical interest, or else that they are the product of an unjust world – something that will not exist in decades to come. I contend, in reply, that WRCs are institutions that animal ethicists can support and which should be of theoretical interest for three different sets of reasons. Briefly outlining these reasons will be useful for situating the institutions here explored within the wider context of animal ethics, especially political approaches to animal ethics – and will reveal why a close analysis of the diets of animals present in these institutions is important.

First, WRCs might be important as refuges for wild animals whose interests have been trampled over – or whose rights have been violated – by humans. Cheryl Abbate (2016; 2020) and Corey Lee Wrenn (2018), for example, explore the ethics of rehabilitation using the example of a lion rescued from a life of abuse in the circus. This makes it sound like WRCs are a kind of 'non-ideal' institution, one that exists only while other harmful institutions persist. Indeed, this is a thought present with both authors. Wrenn, in particular, imagines WRCs vanishing to nothing once the institutions that abuse wild animals within human communities – circuses, zoos, canned hunts, the 'exotic pet' trade, and so forth – have themselves been eliminated. Of course, the elimination of these institutions could mean that WRCs become *more* prominent in the short term. As our societies become more aware of animal rights, survivors of these institutions are going to become more plentiful, as the institutions are dismantled, individuals are rescued, or the institutions 'retire' animals in a show of good faith. So even if WRCs might, as Wrenn seems to imagine, cease to exist in ideal theory, they may nonetheless become more prevalent in the short-to-medium term.

There is another category of wild animals harmed at human hands, however, and this is a category that is more likely to persist more or less indefinitely. These are the animals who have been directly or indirectly but (relatively) inadvertently harmed by human activity. Palmer (2010, 102–6), for example, explores at length an imagined group of coyotes displaced and disrupted by the building of a series of human homes. Even if we may go 'this far and no further' when it comes to the development of land (Donaldson and Kymlicka 2011, 194), *these kinds* of human-wild animal entanglements can become only more prevalent through climate change and other human impacts upon the 'natural' world (for discussion, see Bovenkerk and Keulartz 2016; Schlottmann and Sebo 2018). So, for example, with or without the abolition of intentionally harmful institutions within our societies, what we could call *animal climate refugees* are likely going to exist indefinitely into the future. Given our responsibility for their plight – more on this next chapter – we may have a duty to aid them. WRCs are surely a possible tool for compliance with this duty.

Second, WRCs may be important spaces for aiding animals who are facing suffering and death (including from starvation) for reasons unrelated to humans – as much as such animals can exist in today's world.[2] It could be held that, even though there is no human involvement in or responsibility for the animals' plight, there is an issue of justice here. Martha Nussbaum (2006, 400–1) famously calls for the replacement of the natural world with a just world, while Alasdair Cochrane's cosmopolitan animal ethics presents wild animals as having positive justice claims against us concerning the alleviation of suffering and the staving off of death (Cochrane 2018, chap. 5; cf. Johannsen 2021). Perhaps, then, these theories of justice will necessitate many large WRCs – or something like them (see Milburn 2019b).

Even if we do not take this view – see chapter 7 – many of us hold that it is (in principle) *permissible* to aid free-living animals in need. The extent to which there is a significant moral reason in favour of aiding wild animals is, logically, a different matter, with a range of powerful recent arguments concluding that there are – contrary to common intuitions – very good reasons in favour of aiding wild animals in need (e.g., Hadley 2006; Horta 2018). Reviewing and/or assessing these arguments would take me too far from my current enquiry. What is undeniable, however, is that there are people who desire to help wild animals, and will make every effort to act upon this desire if given the space to do so. Thus, even in an environment where aid to free-living animals harmed by forces beyond human control was not deemed a requirement of justice, privately run WRCs would have a place.

Third, there may be more positive reasons for WRCs to exist. Such reasons are, of course, *in addition* to the negative reasons, but we need not think of rehabilitation centres as *inherently* tragic spaces. In an important article, Donaldson and Kymlicka (2015a) call attention to farmed animal sanctuaries as spaces at the heart of the animal rights movement. They are spaces, they argue, in which alternative relationships with animals can be put into practice and explored (see also Scotton 2017). Rather than victims, or perhaps (potential) ambassadors for farmed animals, the residents of sanctuaries can be thought of as co-community members (in Donaldson and Kymlicka's parlance, co-citizens) and pioneers in the creation of new kinds of communities. Now, Donaldson and Kymlicka's arguments cannot be straightforwardly applied to rehabilitation centres, as different kinds of relationships may well be appropriate with (formerly) free-living animals (or otherwise non-domesticated animals), but it is surely the case that *similar* kinds of arguments can be applied. WRCs might well be, like farmed animal sanctuaries, spaces in which different ways of living can be tried out, and in which different kinds of communities can be forged. These communities and ways of living might (or might not) be very different from those on farmed animal sanctuaries, but that need not make them any less important.

We have, then, three kinds of reasons we should take the institution of the WRC seriously: (1) To aid animals harmed by humans; (2) To aid animals harmed in other ways; and (3) To provide a space for the forging of new kinds of community. However, the institution as it is typically imagined (and practised) is fraught with moral complexity. I wish to introduce what I see as two of the most fundamental complexities – both issues, crucially, directly related to the diets of the resident animals – by way of a reflection on the concept of *rehabilitation*.

There are several different senses in which we use the word *rehabilitation*. One sense concerns a restoration to health. This is the kind of rehabilitation that, in the human context, we might talk about being offered to a sportsperson. An injured athlete is successfully rehabilitated when her injury no longer interferes with her ability to engage in her sport. (This need not necessarily be solely a *physical* injury; we talk of people being rehabilitated after experiencing trauma, for example.) This sense of rehabilitation is obviously relevant for the WRC, as the centre takes in injured or otherwise affected animals.[3] A crucial part of this rehabilitation is going to be feeding. Indeed, feeding may be the primary part of rehabilitation. But this raises an obvious puzzle from the perspective of animal rights: the fact that some of these animals putatively need to eat meat to be healthy. How can we reconcile obligations to feed these animals with obligations to the animals they might eat?

Another sense of the word *rehabilitation* concerns a kind of restoration to moral virtue. This is the kind of rehabilitation that we might talk about in relation to human criminals: at risk of oversimplification, a violent criminal can, in theory, be rehabilitated to become a peaceful member of society through involvement with institutions associated with the criminal justice system. Many of the animals at WRCs are indeed violent. In nature, they will – if successful – kill other animals. Not, of course, because of criminal urges, but *to eat them*. Now, unlike (some) human criminals, they

Table 6.1 Animals who may be helped by WRCs

	Animals relatively difficult to feed without animal products	Animals relatively easy to feed without animal products
Animals likely to inflict violence in the pursuit of food if released	E.g., predatory carnivores, such as big cats	E.g., opportunistic predators who are mostly herbivorous, such as wild pigs
Animals unlikely to inflict violence in the pursuit of food if released	E.g., obligate (thus non-predatory) scavengers of meat, such as vultures	E.g., obligate herbivores, such as rabbits

are not morally responsible for this violence. Indeed, this is *why* they cannot be rehabilitated. It would be a strange – confused, even absurd – WRC that sought to rehabilitate their charges in this sense. That said, given a concern about animals being eaten, we might wish that we could.

A third kind of rehabilitation certainly is the aim of many WRCs. We could describe this as a kind of restoration to 'proper', 'normal', or 'natural' functioning. The aim of many WRCs is to get the animals back to where they 'should' be, doing the things that they 'should' be doing. The ultimate aim, then, is to release these animals back into the wild. The problem is this: whether they are morally responsible or not, predatory animals will (if 'successfully' rehabilitated) go on to inflict death and suffering in the pursuit of food. This, too, raises an ethical dilemma. How can we reconcile the obligations we have to help these animals with the obligations we have towards the animals they might go on to kill?

There are thus two kinds of distinctions that we can make between animals who may be helped in WRCs, and thus two kinds of (sometimes overlapping) ethical problems related to animal diet. First, there is the distinction between animals who are very difficult to feed without animal products (who may be omnivores or carnivores), and those who are relatively easy to feed without animal products (who may be omnivores or herbivores). Second, there is the distinction between those animals who, if 'successfully' released, would likely go on to inflict severe violence upon animals in pursuit of food (predatory animals of various kinds), and those who, if 'successfully' released, would likely not (non-predatory animals, including some carnivores). These distinctions are displayed, with example animals, in Table 6.1 on the previous page.

We can thus see that some sanctuary residents raise both kinds of normative problem, some raise one or the other, and some raise neither.

The Diets of Current Residents

Those animals who can be fed relatively easily on plant-based diets – including both herbivores and relatively herbivorous omnivores – should simply be fed these plant-based diets. It is worth noting that this need not exhaust discussion of their feeding from an animal rights perspective, or especially from a broader food ethics perspective. Consider, for example, the fact that WRCs may be medicating animals. If so, they are likely to be using products that are the result of rights-violating animal testing. Or consider, again, the fact that even plant-based food can be sourced in ways that violate human or animal rights, such as by being sourced from companies that use exploitative labour practices. These are important considerations, but not my present focus.

My present focus is instead those animals who are relatively difficult to feed without animal protein – especially those who are putatively *impossible* to feed without animal protein. Sanctuary personnel face a dilemma with these animals. On the one hand, they want the best for them, and this, it is assumed, means feeding them meat. On the other hand, the reality of harm to other animals and the violation of their rights will – or should – raise serious concerns about feeding meat to residents.

This dilemma was the subject of a scholarly exchange between Abbate (2016; 2020) and Wrenn (2018). Working within Regan's account of animal rights, Abbate explores what can be said about the feeding of a lion – Sophia – who has been rescued from a life of abuse in the circus. Sophia has been physically mutilated, and has no teeth or claws. Nonetheless, 'she must eat other animals if she is to survive' (Abbate 2016, 147). Releasing her is thus not an option – she would starve. Abbate's solution is to propose that Reganites adopt a 'guardianship principle': 'Provided that all those involved are treated with respect, and assuming that no special considerations obtain, a guardian is obligated to harm other innocents when doing so is required to avoid making a vulnerable and dependent victim of injustice at least as *worse-off* as the innocents so harmed are made' (2016, 152, emphasis in the original). This cannot permit animal agriculture, as the animals there are treated as mere resources, and so not treated with respect. Abbate argues, though, that it *can* permit limited hunting of deer (and other animals). So, Sophia can be fed on hunted deer.

Wrenn (2018) objects to Abbate's arguments on several grounds. Ultimately, she argues, Abbate fails to think about the problems she is dealing with systematically, framing them instead as a question for one sanctuary about feeding one lion. When we think about the problem systematically, Wrenn argues, we see that feeding the by-products of factory farming to Sophia will make no appreciable difference for farmed animals – it is corporations and governments, not individuals, who are responsible for the harms to animals caused by factory farming. Meanwhile, Abbate's proposal would seem to create new harms (to deer), and, indeed, further marginalizes already-marginal groups (deer – especially sick, disabled, 'overpopulated' deer). Wrenn thus proposes that it is acceptable to feed Sophia the (purchased) by-products of animal agriculture. In turn, Abbate (2020) defends the idea of thinking – and acting – as individuals, rejects the idea that harm to deer would be *new* harms (given the reality of hunting regulations), and counters that Wrenn's understanding of which animals are 'marginalized' is incomplete.

The debate between Abbate and Wrenn is extremely interesting, but, in my view, it is limited in three key respects. First, both are relying on a very narrow kind of case – that of a victim of direct abuse at human hands – and so their accounts may lack generalizability. Abbate's guardianship principle, for example, seems to apply only in those cases when the animal is a victim of direct harm at human hands. As I have demonstrated, there are a multitude of reasons to be supportive of WRCs beyond this kind of case – and, to repeat, both Abbate and Wrenn have a too-narrow, too-negative view of WRCs as a kind of 'necessary evil'.

Second – and relatedly – neither Abbate nor Wrenn offers a lasting solution. Were Abbate's hunting to become institutionalized, it would involve treating deer populations as a stock of food for carnivorous animals, which would (in her own terms) fail to treat them with respect. Wrenn's proposal, on the other hand, is viable only so long as there *is* a meat industry – to accept this as a full solution would be to commit to the idea that either there will be a meat industry indefinitely, or there will one day be no more WRCs. Both seem problematic.

But third, and most importantly, Abbate and Wrenn are offering justifications for activities that, on the face of it, are rights-violating. It could be that, all things considered, these apparently rights-violating activities are acceptable. But to justify such a claim, one needs to first explore all *other* possibilities. Both Abbate and Wrenn nod towards these (with varying levels of support), but they are by no means their primary focus, which, I hold, they should be. To put it another way, Abbate and Wrenn should not be too quick to accept the idea that 'in order to feed their obligate carnivores, wild animal sanctuaries must, in some way, participate in the intentional harming of other creatures' (Abbate 2020, 171–2).

What are these alternative possibilities? For a start, WRCs should be prepared to explore the possibility that even animals that putatively 'need' meat to survive can thrive on a vegan diet. As discussed in chapter 2, there is research suggesting that domestic cats can survive perfectly well on a vegan diet, even though they are obligate carnivores, and there is no reason to think that we could not craft plant-based diets suitable for other animals, too. It is worth recalling the words of Knight and Leitsberger, from their review of vegan diets for companion animals: cats and dogs, 'and indeed [members of] all species', require particular *nutrients*, not particular *foods*, to thrive (2016, n.p.).

Now, naturally, there has been much research on the nutritional requirements of cats and dogs; no doubt there has been much less on the nutritional requirements of many of the animals who find themselves in a WRC. Indeed, WRCs may sometimes deal with animals of species that are relatively (even completely!) unknown. I also note that WRCs frequently aid animals who are ill, disabled, or injured; very young or very old; or weakened due to environmental strain. These are animals who might be particularly vulnerable to harm if fed a diet very different from the one they have evolved to favour, or even simply if they experience a sudden drastic change in diet. I thus accept that an imperative to simply feed these animals a plant-based diet may be short-sighted, *even if* such an imperative is reasonable in many (or most!) cases when it comes to companion animals.

It is thus crucial for us to revisit the possibility of acquiring animal products without violating the rights of animals. Abbate (2016, 159) calls upon WRCs to be more open to accepting roadkill as a source of meat than they are currently. I would add that WRCs might, themselves, be a source of meat. As other residents die or are dead upon arrival – surely a regular possibility in even the best-run centres – *their* bodies could permissibly become a source of meat. This would not be disrespectful. These are wild animals. We would expect their bodies to be eaten were they to die outside of the centre, and – the whole point of this exploration – their bodies could be used to help another animal survive. (See the related discussion in chapter 2.) Some people may object to this claim. I have argued elsewhere that domesticated animals' dead bodies are owed certain kinds of respect not owed to wild animals' bodies (Milburn 2020a). Part of the reason that the residents' bodies are, comparatively speaking, 'fair game' is precisely that they are the bodies of wild animals and not the bodies of domesticated animals. (Note that I am *not* proposing that we grind up the bodies of companion dogs or our recently deceased human relatives for animals in WRCs.) It is true that animals in WRCs can *come to be* domesticated, and thus become part of our community in much the same way that domestic dogs are. But this is something that the staff at WRCs are keen to avoid, and so it is right to describe *many* animals in WRCs as 'wild', even if they are under a high degree of human control. And this is normatively significant. As we explore new ways to relate to animals in WRCs – as pointed towards later in the chapter – they may become more 'domesticated' and less 'wild', meaning that their bodies could no longer be utilized in this way.[4] But this is all, for now, speculative.

Abbate (2016, 159) is also open to the prospect of cultivated meat as a solution. Wrenn (2018, 165) is surprisingly hostile towards the possibility.[5] Cultivated meat would surely allow a lasting solution to the problem, and could consistently provide high-quality food relatively easily. The wildlife rehabilitation centre, in theory, would not even need to rely on a cultivated meat industry. In a twist on the 'pig in the backyard' model (see Weele and Driessen 2013; cf. Dutkiewicz and

Abrell 2021), we could imagine the centre having facilities to grow meat on-site, using cells acquired from other sanctuary residents. Perhaps the story with which I opened this chapter – rescued pigeons or mice being fed to rescued wildcats – could come true without these rescued animals being killed. Maybe the rescued mice could be harmlessly used to create meat that was then fed to wildcats. It is not obvious that our altruistic rescuer would (or, at least, should) object to this.

We can add that cultivated milk – a related though separate technological possibility to cultivated meat – can be produced in a way completely respectful of animals (Milburn 2018). I do not know whether Wrenn would object to this, but I cannot see why she should. The production of milk without the violation of animals' rights would naturally be very useful for rescue centres handling young animals. And milk could be an important food source for adult animals, as well. We could even imagine milk from a whole range of species being produced – not just the milks humans like to drink. We could produce mouse milk for mice, rabbit milk for rabbits, hedgehog milk for hedgehogs.

And there are other possibilities. For WRCs, unlike – say – individuals who happen to rescue an injured animal, institutional solutions would be desirable. It is hard to imagine that dumpster-diving and backyard chickens (as discussed in chapter 2) would be able to provide all of the animal products necessary – even if, in the world as we find it, animal-based foods that would otherwise go to waste could be redirected to animals in need to lessen the pressure on sanctuaries to support the killing of animals (Abbate 2020, 190). Perhaps, in today's world, WRCs could institute a kind of freeganism by planning with (say) local butchers to take excess meat to feed to the sanctuary residents. This would certainly be preferable to *purchasing* meat. But it is ethically complex nonetheless. Might this partnership help subsidize the butchers, as they will have less waste to deal with? Might it aid the butchers in greenwashing/'humanewashing' their products?

Could WRCs purchase from animal farms without supporting harm to animals? The farming of cows, sheep, chickens, and so on would seem to be out of the question.[6] More likely to be of use are systems of farming non-sentient animals, perhaps including shellfish. As discussed in chapter 2, steps are already being taken to develop food for carnivorous and omnivorous companion animals made from these kinds of animals. The obvious difficulty is in identifying those animals who are non-sentient. Even a little intellectual humility will presumably lead us to conclude that there is a huge range of animals – including many invertebrates – that *may* or *may not* be sentient. We are just not, at present, sure. This means that the farming of invertebrates hardly provides a perfect solution to the present dilemma – even if invertebrates could provide the nutrients necessary to feed the meat-eating animals in the centre, which is not a sure thing. That said, the farming of those invertebrates who *may* be sentient, but are probably not, is surely a preferable alternative to the killing of free-living mammals (à la Abbate) or factory-farmed mammals (à la Wrenn) who certainly are sentient. This would, I propose, be a more suitable 'last resort' than their suggestions (compare Milburn 2015a).

Here, then, is my proposal. Before concluding that they are justified in hunting (as Abbate suggests) or supporting industrial animal agriculture (as Wrenn suggests), WRCS should seriously explore the possibility that their residents – even carnivorous residents – be fed a plant-based diet. If this is not possible, they could feed them plant-based diets supplemented with the bodies of animals who have died naturally or due to accidents – including, perhaps, former residents of the centre – and with food waste. Further animal protein, if it is definitely needed, could be supplied (in the short term) by farmed invertebrates. In the medium term – a genuinely sustainable solution, I suggest – it could be provided by cultured meat and related technologies. Industrial animal agriculture and hunting (when the animals farmed or hunted are rights-bearers) are to be rejected. Given the wealth of alternatives available to WRCS, we have good reason to believe that they are not justified in killing rights-bearing animals to support the animals in their care, *even if* we accept something like Abbate's guardianship principle.

The Diets of Former Residents

As noted earlier, the aim of many WRCS is to release animals once they have been restored to health, so that they may live out their lives in the wild. But this raises an ethical quandary when we remember that many animals are predatory, meaning they will go on to hurt and kill other animals. Now, traditionally, theorists of animal rights have resisted the conclusion that they are obliged to intervene in predator/prey relationships. Regan said that, when it comes to wild animals, we should just 'let them be' (2004, 361) – their rights are not violated by their nonhuman predators, as only moral agents can violate rights, and nonhuman predators are not moral agents (2004, 285). This is correct. But let us not pretend that if the rights of wild prey animals are violated, they are violated by their nonhuman predators. If their rights are violated in the real world, they are violated by humans, or groups made up of humans.

It might be, first, that wild prey animals have positive rights to assistance, and humans (or human institutions) violate these when they fail to assist these animals in their time of need, or fail to protect them when threatened by other animals. Following the relational approach being taken in the present work, the question of whether wild prey animals have positive rights depends on the nature of their relationship with us. I will make no reference to these putative rights here.

But, regardless of their relationship with us, wild animals have negative rights against being made to suffer and being killed. Humans can violate these. Importantly, though, a human does not have to kill an animal with her bare hands to violate that animal's rights. A human who points a gun at an animal and pulls the trigger cannot claim that it was the gun or bullet (both non-agents) that killed the animal, so she did not violate any rights. Nor could she claim this if she simply fired the gun indiscriminately into underbrush, and happened to kill an animal. Similarly, a human could not claim that she was not responsible for the death of an animal if she set a dog on the animal, or even if she let her (violent) dog run into underbrush known to be inhabited by vulnerable prey animals. Nor – I claim – could a human be absolved of responsibility for this dog's actions if the human chose to release the dog into an environment full of vulnerable prey animals.

In short: humans can violate the right of an animal not to be killed even if it is ultimately a dog that kills the animal in question. But if this is true in the case of the dog who happens to be a companion Alsatian, it is also true of the dog who happens to be a wolf who has been saved from death by a WRC. In both cases, the dog, if released into a 'suitable' environment, will kill rights-bearing animals, and the human in question could fairly be held – somewhat, but not necessarily completely – morally culpable. Thus, the rights of the animals killed have been violated.[7]

Readers not convinced of this point are asked to consider the following possibility, borrowed from science fiction. In 'Warhead', a 1999 episode of *Star Trek: Voyager*, the crew rescue a stranded (sentient, intelligent, and morally considerable) artificial intelligence (AI) that has little memory of who/what it is. After beginning to provide it with assistance, they learn that it is a weapon of mass destruction. Nonetheless, the crew decide that the rescued AI has a right to life. The crew attempt to separate the AI from the missile it inhabits, but, once it realizes what is going on, the AI – having regained memory of its mission – demands that it be allowed to continue towards a target containing many rights-bearing beings.[8] The *Voyager* crew are faced with a moral dilemma that is somewhat analogous to that faced by WRC personnel. On the one hand, the *Voyager* crew recognize that it is right for them to rescue the AI, and wrong of them to simply kill it when they see that it will be responsible for harm in the future if properly functioning. On the other – and despite a general aversion to interfering with alien wars[9] – they recognize that putting the AI *back* on the track it was originally on, and where it wants to be, would make them responsible for the deaths of those it killed. This is unconscionable for them.

I contend that the *Voyager* crew are making the morally appropriate decision, here, in refusing to release the AI. If the AI is morally considerable, it is (all else equal) good for them to rescue it, and would be (all else equal) wrong for them to destroy it. However, regardless of the AI's original trajectory, once the *Voyager* crew have 'saved' it, they adopt a degree of responsibility for the deaths that it would cause, and so cannot 'release' it lightly. (The crew recognize this so acutely that they put themselves at grave risk to avoid releasing the AI.) Is it plausible to think that the *Voyager* crew behave appropriately with the AI, but that the same reasoning does not apply to WRCs with free-living animals? In both cases, a morally considerable being is aided, but, if released, will cause significant harm to other beings, leaving those who aided the first being morally responsible for harm.

It cannot be the case that the fact that we are talking about animals being killed, rather than human-like aliens, makes all the difference. Human-like aliens may (or may not) have stronger interests in continued life than many animals, but this does not mean that animals' rights do not exist. Nor can it be the case that the morally relevant difference is the AI's ultimate origin in the hands of moral agents: it is not as if the crew should have changed their action were they to find out that the AI was heading to that planet because of its own malfunctioning programming, or completely randomly, or of its own volition.[10] Nor can it be the case that the fact that the AI has a determinate target makes the difference. If instead of targeting a particular location the AI would hunt through space until it found *someone* to kill, the crew would hardly become justified in releasing it.

A relevant difference may be that the AI is a moral agent, while nonhuman predators are not. It is not clear to me that this can make the moral difference intended. Our intuitive response to the *Voyager* crew's action does not seem to depend on the AI being morally responsible for its intended actions. In fact, the episode is written in a way that presents the AI as *not* morally responsible. A crewmember tells the AI that it has the opportunity to learn, and overcome its original function; until introduced to this possibility, the AI's responsibility is vastly diminished, if extant at all.

If anything, the fact that nonhuman predators are not moral agents makes releasing them even *more* morally problematic, by making those who provide aid to them even *more* morally responsible for the harm they go on to do. There could be no pleading of ignorance, or hope that the predatory animal will 'see the light'. Violence is basically inevitable, and fully foreseeable, and *no one* can be held responsible *but* the agent(s) who release(s) the predatory animal.

We could point out that some animals at WRCs have been directly harmed by humans. Thus, it is not that their first human contact has made them *able* to hunt (as with the AI), but that their first human contact has made them *unable* to hunt. In returning them to nature, we might think, the WRC is simply putting things back as they were, and so no wrong has been done. Imagine if some wild stoats (weasels who predate on rabbits) were rescued from an animal hoarder – could they not be re-released onto a rabbit-filled island, even though this would threaten the rabbits?[11] In such a case, I believe, the rabbits would have a complaint against us. I do think that our society would be *less* responsible for the violation of the rabbits' rights than they would if the stoats had been saved from, say, a (genuinely) natural disaster. This is because the problem that the society helps the stoats overcome is a problem of the society's own creation (more or less). The significance of this is that in some counterfactual world in which the society had not interfered with the stoats, they would still be killing rabbits. Nonetheless, the society would still have *a degree* of responsibility. The society has made (or, more likely, permitted) a decision that sees the stoats released and go on to cause harm.

What about the rescuers, in contrast to society as a whole? I am not convinced that they have *any* less responsibility for the harm to the rabbits in the stoats-rescued-from-hoarder case than in any other.[12] When one rescues an animal, one takes on a degree of responsibility for her. If she goes on to successfully kill other animals because of us, we are responsible for that fact – no matter how we came to be responsible for her. Imagine if the AI that the *Voyager* crew saved was on the planet because of the deliberate interference of some third party, or even the *Voyager* crew themselves. The crew would hardly be absolved of responsibility for the deaths caused by the weapon were they to then repair and 'release' it on the grounds that they were just putting things back to the way they were. Once we become morally entangled with animals who may go on to cause harm, we take on degrees of moral responsibility for the harm they cause. I do wrong in handing a knife to someone about to murder a child to complete some byzantine vendetta, even if I was the one who knocked the knife out of the would-be murderer's hand.[13]

Does this mean that *any* (causal) entanglement with a wild animal makes us responsible for the plight of the animals she kills? I think not. We can make a distinction between, first, *aiding* her in killing other animals and simply being *involved* with her in some way; and, second, between *intending* for her to hunt and *not* intending for her to hunt. Imagine two scenarios in which a human community moves a predatory animal from location A to location B. In the first scenario, they move her from A because A is frequently visited by human children to whom she may be a threat. In the second, she is moved to B because B contains an ample supply of prey animals for her. In the first case, though the human community is causally responsible for *this* rather than *that* prey animal being killed, it does not seem to be *aiding* the predatory animal in her hunting, and nor does it seem to *intend* for her to kill other animals (though that is surely a foreseen consequence). In the second case, however, the community clearly *aids* the predatory animal in having access to prey, and clearly *intends* for the predator to hunt.

Note that the level of entanglement that humans possess with wild animals is (ordinarily) far lower than the level they possess with companion animals. While there might well be a presumption that there is a degree of human responsibility for any harm caused by a companion animal, there is no such presumption for wild animals – even if there has been some prior entanglement between humans and this wild animal (such as a community moving her from one site to another). And even when there *is* human responsibility for harm, this responsibility is surely *much lower* than in paradigm cases involving companion animals. This makes the violation of rights in these cases less serious from the perspective of justice – but a smaller injustice is an injustice nonetheless. Alternatively, as discussed in chapter 4, it might be that the responsibility becomes sufficiently low that the case for injustice is negligible. 'The law does not concern itself with trifles' (*de minimis non curat lex*), as lawyers would say.

Let us return to the WRC. The WRC is clearly more analogous to the community in the second case (releasing an animal in better hunting grounds) than in the first (releasing an animal away from children). Though their primary motivation might be restoration of the predatory animal to health – surely a noble goal – a secondary motivation is restoring the predatory animal's ability to hunt, and, indeed, to 'successfully' release her so that she can go on hunting. This means that the WRC possesses a degree of causal *and moral* responsibility for any harm that the animal would go on to commit after release. The harm that these animals would go on to commit would make the WRC responsible for the violation of animals' most fundamental rights. This would mean that the WRC is, rather than an institution helping animals, an institution committing huge injustices against animals. To put it mildly, this is a problem.

There is one obvious solution: the WRC abandons its goal to release predatory animals after they have been restored to health. As the table earlier in this chapter shows, this does not mean refusing to release all animals, and it need not even mean refusing to release all carnivores. But it does mean that it would refuse to release *many* animals.

What are the alternatives to releasing them? Two options would, I take it, be highly undesirable. First, WRCs could refuse to take on these animals in the first place. This would mean leaving them to suffer and die. However, this response may have its place. WRCs have the ability only to take on certain animals – no matter how much I implored them, a chimp sanctuary would be unlikely to take on a dolphin, even a dolphin in dire need. Similarly, some WRCs might reasonably refuse to take on animals because they do not want to become morally entangled with them. The *Voyager* crew, perhaps, could have permissibly passed by the AI, refusing to get involved, *even though* they could potentially help. Equally, I could permissibly pass by a brawl, not wishing to become involved, *even though* I could potentially help. Once the crew *is* involved with the AI, once I *am* involved with the brawl, the moral facts of the matter change. The same is true with WRCs. While, for many WRCs, this is hardly a perfect solution, refusing to get involved may well be the right course of action in cases in which they cannot avoid responsibility for rights violations if they *do* get involved. One might object that WRCs are *obliged* to help the animal. But this is too demanding. To function, sanctuary-like institutions *must* make difficult choices about which animals to save (Abrell 2021, 125). Sanctuaries frequently refuse to take on an animal if they are not convinced that they can care for the animal in a way respectful to the animal herself; and, I contend, WRCs might reasonably refuse if they are not convinced that they can care for the animal in a way respectful to *other* animals.

Second, the WRC could euthanize the animals. This would at least relieve the animals' suffering, but would presumably conflict with their right not to be killed – especially if their suffering could be alleviated cheaply and easily. It is not, of course, a desirable course for WRCs, as it goes against their central goals.

I thus propose a third alternative. The option that remains is to offer the animals a life in captivity. The predatory animals helped by WRCs should *never* be released.

There is already precedent for this kind of approach. Animals who cannot be successfully rehabilitated to live successfully in the wild are frequently kept by WRCs indefinitely. Indeed, such an approach is standard at some similar institutions. Farmed animal sanctuaries, for example, typically have *no* aspiration to 'release' animals, and will often not even have an intention to 'rehome' animals. Similar is true in the case of sanctuaries focussing on wild animals rescued from the likes of the 'exotic pet' trade – the focus of the exchange between Abbate and Wrenn. So, clearly there is no in-principle objection to refusing to release wild animals from WRCs among WRC personnel. My proposal is simply that we refuse to release them, not for their own sakes, but for the sake of the animals they will (almost) inevitably go on to harm. Releasing animals is contrary not to their own interests, but the interests of others.

This would mean that WRCs, or institutions like them, would become a permanent home for a great many animals. And this, in turn, calls for the place of these animals to be more clearly thought through. Could these animals become members of our community – more like the domesticated animals discussed in chapter 3? How could these institutions be designed to give these animals the space and other resources needed to live a flourishing life – and what is the 'next best option' for those animals who cannot be offered a flourishing life in captivity? How close can or should their relationship with humans be, whether these humans are their caretakers or other visitors to the centre? To what extent, if at all, could these captive animals become a source of income for the centre – thanks, perhaps, to paying visitors? This reimagined institution, then, raises many of the same questions as the farmed animal sanctuary, including those nodded towards earlier in the chapter. It may be the case that the answers will not be the same in the case of farmed animal sanctuaries and the analogous institution with wild animals, but this is going to require careful consideration by animal ethicists and scholars of animal studies. The emerging literature on animal sanctuaries, I contend, should not be limited solely to *farmed* animal sanctuaries.

Concluding Remarks

Ultimately, I offer two proposals concerning WRCs and related institutions. First, such institutions should not be feeding animals meat that is the product of rights-violating practices and industries. This requires some creative thought when it comes to feeding those carnivores who, putatively, need meat to survive. Second, such centres should not be aiming to release those animals who will go on to hunt other rights-bearing animals. If they do so, the blood of these other animals will be, metaphorically, on the centres' hands. In rescuing animals, even if the action is commendable (as it surely is), we can take on a degree of responsibility for their future actions: always causal, sometimes moral. When we bring them back to health, we are (literally) helping them to hunt rights-bearing animals, and this is morally problematic. Refusing to release these animals calls on us to creatively experiment with different kinds of communities – I have not, here, begun to explore what these communities might look like, but have pointed towards the emerging literature on farmed animal sanctuaries as beginning that conversation.

I hope it is clear that my position is not 'anti-nature'. It might be contended that condemning animal-on-animal violence sounds a long way from endorsing (even with regret) the 'natural functioning' of predator-prey relationships. But the point I am making is precisely that the predatory actions of released animals are *not* natural, but (at least partially) anthropogenic. Concern for the actions of the staff of WRCs is completely consistent with offering no objection to *genuinely* natural acts of predation.

I also hope that it is clear that my position is not 'anti-rehabilitation'. My position is simply that WRCs, like all spaces/institutions involving animals, need to transform to genuinely be respectful of animal rights. That kind of transformation may be difficult. But I am certainly not, here, intending to offer a kind of utopic ideal theory. On the contrary, I have offered practical solutions to both problems identified, though accept that there is further work to be done. And let us not forget that these institutions are *already* devoted to the worth of individual animals. It is surely not that big an ask, conceptually speaking, to call on them to extend the respect that they offer to their residents to those animals who would be killed for, or by, their residents. Bizarre though they may sound in the abstract, ultimately, this is all that my proposals amount to.

7

Animal Strangers

The idea of reducing animal suffering is not a particularly strange one. But the idea that we should be trying to reduce the suffering faced by animals in nature, in the normal course of their lives, is one frequently met with incredulity. But this is precisely what some philosophers are prepared to claim. Martha Nussbaum writes that we should support the 'supplanting of the natural by the just' (2006, 399–400). Meanwhile, Jeff McMahan, in an op-ed in the *New York Times* – as well as more scholarly work (2015) – embraces the 'heretical conclusion' that it would be 'good if predatory animal species were to become extinct and be replaced by new herbivorous species,' because of the harms they inflict upon prey (2010). And the position is one increasingly being taken up by animal activists. The Wild Animal Initiative is a charitable group committed to making the lives of wild animals better, and was recognized as one of four 'top' charities by Animal Charity Evaluators in 2020. While it may sound bizarre – even confused, quixotic, absurd – that we should be trying to take steps to stop wild animals from becoming (through no fault of ours) ill or injured, to stop them from starving, or even to stop them from being predated, this is precisely what some animal ethicists and activists are now exploring.

Wild animal suffering is, quite clearly, a concern for anyone interested in the ethics of feeding animals. Two of the major sources of animal suffering in nature are, first, predation, and, second, starvation. If predators did not eat prey, and if starving animals were provided with food, much wild animal suffering would be alleviated – though there would still be difficult questions about how we could go about achieving this, and what the alternative would look like.

Let us be clear what a concern with wild animal suffering is *not* about. It is *not* about the conservation of species or ecosystems – or, not directly.¹ And it is not about protecting wild animals from the aggression of humans or domesticated animals – or, not primarily. Instead, it is about recognizing that the suffering and death of animals is a bad thing, and thus something that we have some reason to work against, *even if* that suffering and death is nothing to do with us, and *even if* that suffering and death is wholly 'natural'. And, as I use the phrase here, it is not about harms to those almost-wild or formerly wild animals that have been discussed in previous chapters. It is not about harm to animal neighbours, including those who live in our towns and gardens, or on our farmland. Nor is it about harms to wild animals who now live with humans in some way, such as those in wildlife rehabilitation centres.

This book started with the idea that animals have key negative rights based upon their interests. In normal circumstances, this rules out many of the direct harms that humans inflict upon wild animals, such as hunting, trapping, and fishing. We have also seen that we can have something that looks like a positive duty towards wild animals when it comes to the harms that our companion animals would inflict upon them. *We* are responsible – to a degree – if domesticated animals in our care hunt, kill, or otherwise harm sentient wild animals. Consequently, the ethical framework advanced in this book so far has, if you like, encouraged a mostly hands-off approach to wild animals – it has encouraged us to 'let them be' (Regan 2004, 361), and encouraged us to ensure that our companions let them be. This conclusion matches our intuitive assumptions about our duties to wild animals – what Clare Palmer (2010) calls the laissez-faire intuition.

On the other hand, the framework here developed has not ruled out *positive* interaction – interaction that could aid the animals in some way. For example, one of the reasons (I argued) that it is worth taking wildlife rehabilitation centres seriously is precisely because they are (or could be) important institutions for aiding individual wild animals. Some advocates of the laissez-faire intuition are prepared to challenge these sorts of positive interactions. Advocates of what Palmer calls the *strong* laissez-faire intuition will say that we have a duty to respect 'nature' by leaving it alone. This is not the sort of challenge that will be explored in this chapter, as it seems to rest upon a valorization of nature and natural collectives (species, ecosystems, and so on) that starts from a very different place than normative frameworks that take seriously the value of individual animals (Palmer 2010; Johannsen 2021, chap. 2). That said, it should be clear that opponents of intervention who appeal to this kind of environmental logic should be excited about the prospect of an internal critique of intervention – something that, to a certain extent, I will be offering. Such critique presents the chance for a reconciliation, of sorts, between environmental and animal ethics.

Rather than focus on the environmental challenge to intervention, this chapter asks whether we have reasons grounded in the rights of individuals to support or oppose intervention. The focus will be theoretical, rather than practical. The idea that we are permitted (or, stronger, have a *duty*) to massively interfere with food relations among wild animals will be bizarre to many readers, and so we need to ask the more foundational questions – in time, this will lead towards progress on the more practical questions. As a result, this could fairly be called the most speculative chapter in the book.

Of course, impracticality may *itself* be an objection to interfering in nature. But to accept that we should not be interfering because it is impractical – that is, we should not intervene because we do not know how to do so effectively – is itself controversial. To say that we should not intervene because we are unsure of how to do so seems to presuppose the idea that if we *could* intervene, we should (Milburn 2015b, 275–8). And that is something that many will want to deny.

This chapter will advance as follows. First, I will explore the *prima facie* case for aiding wild animals in need, and identify two subtly but importantly different ways of framing the conclusion that we should be helping them: that it would be *good* if we helped them, and that we are *required* to help them. Though I remain neutral on whether we have such a duty, I propose that we explore the possibility that we have a duty of beneficence to aid wild animals in need, which is importantly weaker than the claim that we have a duty of *justice* to aid them.

Second, I will turn to explore two food-related duties we might think we have concerning wild animals: the duty to stop them from becoming food, and the duty to provide them with food. I will argue that a duty to stop animals from becoming food encounters problems when we consider the negative rights of animals. In short, it is hard to see how we could realize this *positive* duty while respecting *negative* duties. This means that even if we have (in theory) a positive duty to aid wild animals – a possibility on which I remain neutral – we might have every reason to oppose many particular interventions, and especially the prospect of wide-scale intervention. However – turning to the question of feeding wild animals – I will argue that, given the reality of climate change and the consequent struggles that wild animals face in acquiring food, it is increasingly difficult to maintain that we are not entangled with them in morally salient ways. Thus, even on the kind of relational framework advanced in this work (which is, on the face of it, anti-intervention) we may find that we frequently have obligations to aid wild animals in need.

The *Prima Facie* Case for Intervention

Nature is not a happy place for animals. Without human involvement – and barring catastrophe – populations survive, but frequently at the expense of individual animals. This point is worth stressing: the flourishing of populations frequently occurs not because of the flourishing of individuals, but despite the suffering of individuals. This can be despite the suffering of members of other species. For a single sparrowhawk to survive for a year, many dozens of birds must die a horrific death in her talons – and many more may experience the terror of the chase, be injured but survive (perhaps to die a lingering death), or starve when their parents stop returning to their nest. But it can also be despite the suffering of members of the same species that is, as a whole, 'flourishing'. A blackbird will lay a clutch of three to five eggs, and will have two to four clutches a year. Blackbirds can live for over ten years, meaning that a particularly successful individual could have scores of young over her lifetime. But if the population of blackbirds is remaining constant, then (statistically) only two of these (one per parent) will survive to breeding age. The others will die when young – disease, parasitism, predation, starvation, exposure, and accidents will claim these animals. And, let us be clear: blackbird parents are relatively attentive, with the young fed and protected for the first several weeks of their life. The odds of a newborn blackbird chick making it to adulthood are relatively good when compared with the odds of newborns of many other species making it to adulthood.

Individual wild animals die from disease, starvation, predation, and exposure; they are afflicted with parasites, crippled by injuries, and locked into literal mortal combat with those they would eat, those who would eat them, and rivals competing for territory, mates, or food. And while the phrase 'wild animal' may evoke charismatic megafauna of various kinds – elk, eagle, elephant – the overwhelming majority of sentient animals in the world are smaller. On conservative readings, they are fish, rodents, reptiles, and amphibians. (On more liberal interpretations of which animals are sentient, they are crustaceans, or insects.) The taxa made up of smaller animals have, on the whole, evolved reproductive strategies that allow for the mass suffering and death of their young. Recall that, in a stable population, an average of only one offspring per parent will reach breeding age. The problem is that the majority of animals[2] favour a biological strategy of having many more young than this, with parents investing comparatively less energy in the survival of each offspring. Oscar Horta uses the example of cod, who may lay tens of millions of eggs, and sunfish, who may lay *hundreds* of millions of eggs (2017, 266). Some of these eggs, of course, do not hatch – but many do. And if only two offspring are likely to make it to adulthood, what does that mean for the rest of these animals? It means that they die while young, and perhaps die in very painful ways. Kyle Johannsen uses the example of meadow voles. These animals do not have so many young as do cod or sunfish, but they are still examples of so-called *r*-strategists. (*K*-strategist animals, on the other hand, have comparatively few young, but invest considerable energy into helping them survive. Whales, bears, and humans are *K*-strategists.) Drawing upon one study as indicative, Johannsen observes that 88 per cent of meadow voles die in their first month of life, and the majority of these animals are victims of predation – far from a suffering-free death (2021, 13).

The number of wild terrestrial vertebrates on earth at any one time is at least in the hundreds of billions, with wild marine vertebrates at least in the tens of trillions (Tomasik 2019). Many of these animals will be sentient. And many – most – of them will belong to *r*-strategist species. (Invertebrates, at least some of whom are likely to be or may be sentient, are far, far more numerous, and much more likely to belong to *r*-strategist species.) And even those who do not belong to *r*-strategist species – like many large mammals – will face a great deal of hardship in their lives. This hardship matters. Animals have an interest in not suffering, and have an interest in experiencing positive things. (We might be reluctant to ascribe an interest in continued life to wild animals whose lives are likely to contain a disproportionate amount of suffering.) And we *could* help them. It would not be easy, and we might be reluctant to endorse widespread aid to wild animals out of a fear of creating more suffering through inept (if well-intentioned) interventions. But with good will, careful research, and the investment of time and money, we could collectively come up with ways to make animals much better off (Johannsen 2021). We might start (relatively) small, ensuring steady supplies of food for animals in winter, inoculating against common diseases, or favouring ecosystems that happen to contain less suffering over those that happen to contain more when we are rewilding spaces.

Over time, we could scale up, and – without any exaggeration intended – reshape nature into a space in which animals live better lives. What might this look like? A range of options are possible (Milburn 2019b). Perhaps wild spaces would start to look more like a very animal-friendly version of a zoo, with animals who might do harm to each other separated, and food and care provided to inhabitants (directly or indirectly) by humans. Or perhaps we could rewrite the genetic code of animals so that they are immune to diseases, or do not need to hunt other animals, or have far fewer young, or do not feel pain. Or maybe mass contraceptive drives (or wide-scale habitat destruction) could radically reduce the number of wild animals – fewer animals means, all else equal, less suffering. Or maybe we could have some combination of these approaches, or something else altogether. My point is that we could, in theory, do something.

But should we start on this journey? From the perspective of a consequentialist ethic (or, indeed, any ethical theory that weighs consequences heavily), the question does not seem difficult. The suffering in nature is gargantuan. Even if it is going to be difficult to do something about it, it will likely be worthwhile trying.

But, though the ultimate goal of eliminating suffering may seem almost overwhelmingly difficult, the first steps mentioned above are achievable to the extent that they are already practised. Not-for-profit groups, state-sponsored groups, and individuals *already* engage in the inoculation of wild animals (often to protect humans or domesticated animals), *already* feed animals in need (perhaps because they are members of endangered species, or because humans like to watch these animals in particular places), and *already* make informed choices about which kinds of habitats to create or ecosystems to foster. Now, they likely do these things primarily for reasons other than reducing wild animal suffering, but it still shows what can be done. Consequentialists (and others) thus increasingly support intervention in nature (Sebo 2021), and even if they do not support it in practice, they should be open to it *in theory* (Milburn 2015b).

But for advocates of animal rights, the question is more complicated. There are at least two possibilities. The first is that animals have a right to assistance in the case of overwhelming suffering. Alasdair Cochrane (2019) is one rights theorist who is open to the idea that we should be assisting wild animals in need. However, the obligations that this would impose upon us (as political communities) would be overwhelming, and Cochrane stops short of claiming that actually existing wild animals all have such rights. Instead, he says that the potential rights of these wild animals must be moderated by a principle of proportionality, meaning that – in practice in the world right now – many animals do not have a right to assistance, given the limits of our ability to aid animals suffering in nature. (To adopt the jargon of interest-based rights: on this picture, wild animals always have a *prima facie* right to assistance, but they typically lack, in the real world as we find it today, a concrete right to assistance.) But, nonetheless, Cochrane opens the door to the deeply radical claim that wild animals have a *right* to our assistance.

The framework deployed in this book proposes that the positive rights of animals are best understood relative to our relationships with them. On the face of it – and though this will be challenged later in the chapter – it is hard to see why wild animals would have a positive right to assistance. We lack personal connections to these animals. They do not have dependency on us – either an internal dependency or an external dependency. We are not causally or morally responsible for their plight. Indeed, there may be a great deal of mutual ignorance. No humans know that (say) this particular frog lives deep in the Amazon rainforest, and that frog has never encountered a human – or even (perhaps) anything anthropogenic. Nothing about the (lack of) relationship we have with wild animals seems to suggest that we owe them duties of assistance.[3]

The other possibility, championed by Johannsen (2021, chap. 3), is that the duty to intervene to help these animals is a duty of beneficence. That is, the moral propriety of intervention does not arise from justice or the rights of animals, but rather from our more general responsibility to alleviate suffering where we are able. A duty of beneficence is still a *duty*, but, to use the language of deontology (literally, the science of duty), it is an *imperfect* duty. This means that we must be beneficent, but not that we are obliged to be beneficent towards everyone at every opportunity that arises. Nonetheless, the obligation on us could be very strong. To use an apposite example, a duty of beneficence might not obligate us to give away all our savings – but it may still obligate us to give away half our savings.

Imperfect duties contrast with perfect duties, such as respecting rights. We must (almost) always adhere to our perfect duties, even when – this is crucial – doing so might make it more difficult for us to carry out our duties of beneficence. So, for example, I am not allowed to raid my neighbour's herb garden when her thyme and sage would really improve the stew I am making to feed the homeless. The utilitarian might be open to such a raid, but the point of rights is that they are constraints on the actions that we can take, whether in the pursuit of selfish *or* noble ends. Indeed, even in the case when my neighbour's carrots and potatoes would add bulk to a frankly inadequate stew to feed those in need, I am not allowed to raid her vegetable patch. Only in the case of, to borrow an evocative phrase, *catastrophic moral horror* (Nozick 1974, 30) are we permitted to disrespect another's rights to carry out our duties of beneficence. Perhaps, for example, if my neighbour's orchard was overflowing with produce, and was the only nearby source of food, and someone at my door was literally on the verge of death from starvation, I could be permitted to steal a few apples.

The idea that we have an (imperfect!) duty of beneficence to aid wild animals in dire need has a lot in its favour. It is good to alleviate suffering – even when the other party has no right entailing that this suffering be alleviated. It is good to provide opportunities for positive experiences, even when the other party has no right to be provided with these opportunities. And the idea that we would have no duty to assist in cases of catastrophic *human* suffering is almost unconscionable. If we are consistent anti-speciesists – as we should be – then, on the face of it, we should be ready to extend that same conclusion to cases of *animal* suffering.

Despite this, I am not going to commit to the claim that we have a duty of beneficence to aid wild animals. I am open to it, and I am interested in exploring the consequences of such a duty. But I allow that there is a range of ways that such a duty could be resisted.[4] Instead, what I am interested in exploring is the idea that, even if we have a duty of beneficence, there are important limits to what we can do in its pursuit. To explore this, let us turn to a first way that we could help wild animals – we could help stop them from *becoming* food, and protect them from their (wild, nonhuman) predators.

The Predator Problem

The idea that animal rights (understood broadly) commits us to preventing predators from harming prey has been at the core of both stark defences of intervention in nature (e.g., McMahan 2010) and of arguments seeking to portray animal rights as absurd (e.g., Sagoff 1984). Let's take a lion's hunting of a gazelle as an example. To echo claims made in earlier chapters, it is *not* the case that lions violate the rights of gazelles whom they hunt or otherwise harm. This is because lions are not moral agents. They have no conception of right or wrong, and they lack the capacity to reflect upon the morality of their actions. They thus lack duties to respect rights. Equally, though I generally have a right to not be struck in the face, it is not the case that my right is violated if a baby, while being cradled in my arms, swings her closed fist into my jaw. Thus, a gazelle's negative rights can be sufficient to make it illegitimate for a (so-called) sportsperson to hunt her, but not enough to make it 'illegitimate' for a lion to hunt her. Indeed, the latter claim makes little sense. For the free-living gazelle to be the victim of a wrong when she is hunted by a free-living lion, we need to posit a positive duty on the part of human agents – hence the focus of the previous subsection. As argued in chapter 3, there are strong positive obligations when it comes to companion animals (or, for that matter, human babies). But it is unclear that there are when it comes to gazelles.

Let us imagine, for the sake of argument, that we *do* possess the positive duty of beneficence presented above. Would this mean that we are obliged to save the gazelle? The first thing to note is that a duty of beneficence, as an imperfect duty, does not oblige us to save *all* gazelles from *all* lions. But, it might be pressed, we would have a duty to save *this* gazelle from *this* lion, if it was *this* gazelle and *this* lion in front of us. (I have no duty to patrol the streets looking for toddlers about to step out into the road – but if one steps out in front of me, we might think, I have a duty of beneficence to pull her back.)

The second point to note is that a duty of beneficence cannot call upon us to do that which we are unable to do, and it cannot call upon us to do that which puts us at dire risk.[5] If someone walking on the savannah happens across a gazelle being chased by a lion, it might not be obvious that she can do anything to save the gazelle, or, at least, that she cannot do anything without risking that she herself is attacked by the lion. If either of these things are the case, then she surely cannot have a duty (of beneficence or otherwise) to save the gazelle. Of course, there are lots of cases in which intervention could be both feasible and risk-free – most predatory animals are sufficiently less threatening than a lion. And if she happened to be an experienced rifleperson and be carrying a high-powered rifle, maybe she could do something in this case, too.

A third consideration is, for present purposes, the most significant. A duty of beneficence is (generally) overridden by duties of justice to respect rights. If the only way that our walker can realistically expect to save the gazelle is to use her high-powered rifle to shoot the lion, then it is surely wrong that she has a duty to save the gazelle. The lion has a right not be shot at, deriving from her key negative rights. This is a duty of justice.[6]

This is not to say, incidentally, that the use of violence against attackers is always illegitimate. If an aggressive individual attacks my dog while I am walking in a public place, it may well be legitimate for me to visit violence on this individual to protect my dog. This is because the violent individual is (we can assume) a moral agent, and thus morally responsible for her attack. Though she has rights not to have violence visited upon her, she waives these when she attacks a dog without good reason.[7] But what if she was *not* a moral agent? What if the attacker was – say – a predatory wild animal? A key difference between my dog and our gazelle, here, is the particular relationship I have with my dog – in all kinds of ways (moral, causal, affective, and more) I am tied up with my dog, and my dog's being where she is. I thus have very particular *special* duties concerning my dog, as argued in chapter 3. But we see now that the question about protecting my dog from a predatory animal takes on a different form from the question about protecting a wild gazelle from a predatory animal. My dog has (what we might want to call) certain positive rights relative to me (and, indeed, others) that the gazelle does not. And when it comes to a clash of my dog's rights against the predator's rights, the predator may lose out. Even if I am not *normally* permitted to give a fox a swift kick, if she is coming for my small dog, that is potentially an ethically appropriate action. And even if I am not *normally* permitted to shoot a lion with a high-powered rifle, if she is coming for my dog, that is potentially an ethically appropriate action.

My dog's rights will not always win out, however. My special responsibilities towards her do not justify me taking just *any* action to defend her. My dog and a fox might be relatively well-matched, but a fox is no match for me – in almost any likely violent encounter between my dog and a fox, it would be disproportionate for me to *kill* the fox, given that I could get her away putting myself at only minimal risk. And it would be disproportionate for me to kill a predator – let us now imagine a lion – in a particularly cruel way *if there were some quicker means at my disposal*. (I should put several bullets into her quickly, not blow off her leg and leave her to bleed.) And we can even imagine examples in which my responsibilities to my dog are outweighed by responsibilities to her attackers meaning that, tragically, the right thing to do is to leave my dog to be attacked. I cannot throw a grenade into a crowd of human ten-year-olds, even if they are throwing rocks at my dog and (for some reason) my only chance of preventing the attack is by blowing them up.

Where does this leave us? Advocates of protecting prey from predators might reasonably say that there are many cases in which we could protect prey from predators without violating the rights of the predators. Predatory animals do not have a right to any particular prey animal, and, even if it might be better, all else equal, that we do not disturb and disrupt them, they surely do not have a *right* not to be disturbed. Thus, for example, if we come across a wildcat approaching a nest of peeping waterfowl chicks, we could permissibly spook the cat, encouraging her away from the nest. Though it would hardly be virtuous to spook her just because we would enjoy seeing her run, spooking her for the chicks' sake could not be viewed in the same terms. Such an action seems to be a permissible way to carry out an imperfect duty of beneficence towards the chicks. There may be other possible approaches.

But it is one thing talking about spooking *one* wildcat approaching *one* nest. (Or of one nest of waterfowl being moved to a sanctuary, or what have you.) It is quite another thing to talk about institutionalization of this sort of protection. Putting aside questions of feasibility and practicality, we can see that if we were to somehow *always* spook wildcats (and more) *whenever* they approached nests (or similar), we would be acting in a way that was quite significantly contrary to the interests of wildcats (and other predators). These animals would lose basically any opportunity to feed, meaning they will lose the opportunity of simple pleasures (perhaps some of the only positive experiences realistically open to them), and will starve, experiencing suffering and an early death. The same can be said for any dependent young. This means that while one spooking might respect the cat's rights, many spookings will not.

So, if we were to move to some more institutionalized means of protecting prey from predators, we would need to work out some way of feeding the predators – or of taking some step to ensure they do not need to eat. And thus we start to arrive at the kinds of proposals alluded to earlier. Reforming nature in the image of a (kind of) zoo; massively controlling wild animal populations; genetically modifying wild animals to ensure that their harmful habits are constrained; and so on.

But perhaps these activities, too, violate the rights of animals. And recall that, on a vision of rights as side constraints on the actions that we may take to pursue good ends, the rights of animals will prevent us from pursuing duties of beneficence towards other animals.

Any attempt to massively reduce predation that makes some effort to respect the rights of predatory animals is going to involve a great deal of restructuring of wild spaces and wild ecosystems. This may be a literal restructuring: a building of fences to keep predators away from prey, a (re)engineering of habitats towards predator-free ecosystems, or similar. But it may also be a figurative restructuring – genetically modifying predators so that they are no longer predators, for example. Either way, there will be a great deal of human presence in animals' spaces and lives, a great deal of human control over animals' spaces and lives, and a great deal of disruption of animals' spaces and lives. Could these things constitute a violation of animals' rights in and of themselves?

Let us put this another way. As noted, the kinds of reasons that we have to support human control over certain animals' diets (explored over previous chapters of this book) do not hold when it comes to wild animals who live free from humans. But, what is more, our particular relationship with these animals might indicate that we have good reason *not* to interfere with their diets. These animals have perhaps had *no* (direct) contact with us, in which case – given the horrific history of human-animal relationships and mirroring the way political communities continue to protect members of 'uncontacted' human groups – we should surely take it that the default position would be one of non-interference. Or else these animals have had *some* contact with us, but 'voted with their feet' (Donaldson and Kymlicka 2011, 66), choosing to remain separate. Again, the default position in such a case would surely be one of non-interference.

Based on both their particular interests and their particular historical entanglements with us, we might think that these animals – predator *and* prey – have an interest in being left to control their own fate. This interest could be expressed in a variety of ways: property rights, sovereignty rights, or simply habitat rights, for instance.[8] The idea in each case is broadly the same (Milburn 2016c). To realize the goods open to these animals, to live the particular form of life they wish to live, to have a chance at the life of the kind that is unique to them, these animals require that their space is respected – both in the sense that it remains *their* space, and in the sense that it continues to contain the things that they need (relatively) unspoilt.⁹ Though it is surely permissible – perhaps commendable, as a duty of beneficence – to save particular animals we encounter from grisly deaths in the jaws of predators, it is hard to imagine that any attempt to realize this on a large scale could be consistent with respecting the territorial rights of animals.

What options are left for the interventionist? First, they could acknowledge that intervention to protect prey from predators must be respectful of the kinds of territorial rights just noted. This is precisely the route taken by Johannsen, who allows that the property rights that ('it is plausible to think') animals possess place serious limits on the kinds of things that can be done in pursuit of alleviating wild animal suffering (2021, 58).¹⁰ So, Johannsen thinks, the intentional destruction of natural environments is ruled out, for failing to respect animals' territorial rights. But – Johannsen argues – forms of genetic editing remain permissible. For example, predators could be genetically edited so that they are no longer predators, or prey could be genetically edited so that they are not harmed by predation (e.g., edited to be no longer sentient).

Second, the interventionist could return to the radical idea that wild animals have a *right* to assistance in the case of predation – as noted, this is unlikely to hold up on a theoretical framework that takes relationships very seriously, like the one advanced in this book.

Third, they could argue that the plight of animals in nature represents the kind of *catastrophic moral horror* that justifies overriding the territorial rights of animals. Given the scale of wild animals' suffering, it might seem like this third case is plausible. It would mean that *even though* the duty to protect animals from predators is a duty of (mere) beneficence – recall that this duty is being granted for the sake of argument – rights could be overridden in pursuit of its being realized. But if that is so, then we should engage in the minimum overriding of rights possible.

Elsewhere (Milburn 2016c), I have explored a similar idea. We could try to find a way to intervene in nature that is nonetheless respectful of the kinds of interests that accounts of sovereignty rights, property rights, or territory rights are instituted to protect. Let us focus on sovereignty rights, as conceived by Donaldson and Kymlicka, as an example. Sovereignty is important, they say, because 'sovereignty rights – like indeed all rights – should be understood as protecting certain important interests against certain standard threats. In this case, sovereignty protects interests in maintaining valued forms of social organization tied to a particular territory against the threat of conquest, colonization, displacement and alien rule. This moral purpose, we argue, is equally applicable to humans and to wild animals. Indeed, animals arguably have even stronger interests in maintaining these territorially-specific modes of organization, since they are often more dependent on specific ecological niches' (2013a, 151–2).[11]

Sovereignty, then, is a tool used to protect animals' interests; it is not an end in itself (cf. Cooke 2015; Ladwig 2015). Assistance for wild animals at risk of being predated could (in theory!) be structured so that it did not involve and was not a screen for conquest, colonization, domination, human settling, or human 'development'. (And, of course, intervention is being proposed in this case precisely to *prevent* conquest and domination – animal conquest and domination of other animals, at least.) It could also be structured to minimize disruption and displacement of animals. The question would become not how we could prevent intervention so as to protect animals' territorial rights; it would be how we could design intervention so that our duties of beneficence could be realized without undue disruption to animals.

Again, Johannsen takes crucial first steps in this direction. Although allowing that even the most animal-friendly route towards alleviating wild-animal suffering on a massive scale will involve the violation of some animals' rights, he argues that some routes involve fewer rights violations than others. So, for example, he argues that while genetic editing would not involve the violation of animals' property rights, it would necessitate animal testing. But testing on *non-sentient* animals would not be rights-violating, and even if testing was required on sentient animals, the massive harm (or, though this is not his phrase, catastrophic moral horror) of suffering in nature may justify it (Johannsen 2021, 69–72).

We have thus reached the surprising conclusion that, *in theory*, protecting animals from their predators could be a legitimate and realizable duty of beneficence. This is true even if such protection takes place on a wide scale, even if we reject the idea of positive rights for wild animals, and even if we take seriously the territorial rights of wild animals.

But I stop short of offering a full endorsement of intervention to protect prey from predators, for three reasons. First, I again stress that I have not offered a rigorous argument for, or even endorsed, the duty of beneficence described. Instead, I have explored what our duty might be if we take it for granted. Second, I have not offered anything close to a full proposal for what this kind of intervention might look like. At best, I can echo Johannsen's call for more research on the possibility. And, third, if this is a duty of beneficence, then it is not a duty of justice, and thus not a duty of the state. Individuals, charitable groups, and so forth might seek to make (their patch of) 'nature' more animal-friendly – but they have no business forcing others to aid their enterprise. It is not, for example, the kind of enterprise that justifies massive taxation to support.

Duty to Feed

Let us turn to our second question. This is not a question of a duty to protect prey animals, to stop them from becoming food, but the more straightforward (and, as we have seen, not unrelated) question of whether we are obliged to feed wild animals in need. After all, many animals suffer and die because of a lack of access to food, or are forced to put themselves in mortal danger to seek food out. And, as just explored, animals harm other animals in pursuit of food. A great deal of suffering and death could be alleviated if we fed wild animals. So, should we feed wild animals in need?

On the face of it, my answer to this question should be clear from what has been said so far. Based on the kind of relationship we have with these animals, they (apparently) do not have a *right* to be fed. We may have a duty of beneficence to aid them, but realizing a duty of beneficence (generally) has a lower moral priority than respecting a right. What is more – as explored in the previous section – we must take every step, when carrying out this duty of beneficence, to eliminate (or, at the very least, minimize) the violation of the rights of animals, including any territorial rights possessed by wild animals. With predation, this rules out (for example) undue interference with predatory animals – we cannot shoot them. When it comes to providing food to needy wild animals, we must ask questions about the source of this food. We cannot feed these animals with food that was itself acquired in rights-violating ways, meaning that (for example) we cannot farm sentient animals to kill so that their bodies can be distributed to wild animals. And, no matter how we source food for wild animals, we must be careful that our interventions do not unduly pull control of wild spaces from animals, do not serve to cover imperialistic intentions of supposedly altruistic humans, and so on.

This might seem to be the answer that should be favoured. But, in fact, matters are not so simple. The core premise of the claim that animals do not have a right to be fed is that we have had little or no impact on their lives – we and they are not intertwined in any of the ways that we are intertwined with (say) companion animals. But this point is vulnerable to challenge. We today live in what some commentators have called the 'Anthropocene'. This is an epoch in which human influence on the natural world is both pervasive and defining. But if human influence on the natural world – including the animals who inhabit it – is pervasive and defining, then we cannot help ourselves to the assumption that the lives of wild animals and the lives of humans are unentangled. While some wild animals may be unaware of us and we may be unaware of them, if we are entangled with them in chains of causal (and moral) responsibility, then we may have obligations to assist them.

Particularly important to thinking about the Anthropocene are the effects of anthropogenic climate change. The chemical makeup and temperature of the earth's air, water, and soil – even in comparatively 'untouched' parts of the world – bear the marks of human activity. This impacts the animals who inhabit these places. Sometimes, that impact will be direct. Some animals are poisoned by polluted food or die from thirst or hunger in response to drought. Others are impacted much more indirectly. Chemical and temperature changes impact the kinds of plants, fungi, and microorganisms present in ecosystems, which in turn has an impact – which may or may not be negative – on those animals who consume them, and on the animals who consume *them*, and so on. Collectively, humans are thus causally responsible for the plight of many animals who are suffering and dying because of anthropogenic climate change. But this responsibility is not *merely* causal. As we are aware of the impacts of anthropogenic climate change, yet collectively make only minimal changes to our lives – even though we *could* make larger changes, if only we had the will – we are (collectively) *morally* responsible for this suffering and death, too. Or, at least, partially so.

But if we are both causally and morally responsible for the fact that some animals are unable to find food, the case for non-intervention is thrown into disarray. Though we might not have any personal or causal relationship with the animals themselves – unlike the personal and causal relationship we have with our companions – we must shoulder responsibilities in response to the harms we have done. And we do not need to assume that wild animals possess positive rights to assistance, or even that we have a duty of beneficence to aid them, to reach this conclusion.

Take, for example, the case of polar bears (Palmer 2021). Due to the impacts of global warming, ice patterns in polar bear habitats have changed. This has meant that polar bears have had significantly less access to seals, their favoured food. In turn, this has encouraged polar bears to search for food closer to human settlements. Because of the threat that polar bears present to humans, the hand of the relevant authorities has been forced, and so simply ignoring the polar bears' plight has ceased to be an option. Conceivable responses to the polar bears include relocating them, killing them, or providing supplemental feeding to help them survive.

Like all real-world cases, there are a range of factors that need to be considered, here, and concerns with practicalities will mean that neat solutions may evade us. For example, commercially available polar bear kibble contains animal products, presumably sourced from exploitative, rights-violating industries. For the purposes of illustration, however, let us put this point aside, and imagine that a plant-based (or otherwise rights-respecting) kibble was available that would suit polar bears.[12] If we do this, we can see that supplemental feeding is by far the preferable option. Killing the bears is obviously rights-violating.[13] Genuine euthanasia of suffering bears may be permissible, but describing the killing of these starving bears as *euthanasia* is doubly egregious: not only could we help the polar bears, but we are responsible for their plight. (It would be a bit like me 'euthanizing' a cyclist, even though it was *I* who carelessly hit her with the car and *I* who could keep her alive for now by providing first aid. For more, see Lorenzini 2020.) Relocation, meanwhile, is deeply disruptive for the polar bears, bad (even, perhaps, rights-violating) for the seals at the new location, expensive, and may not even be a practical solution at all, given the particularities the scenario. Providing supplemental feeding – even if it might be somewhat inconvenient or expensive compared to shooting the bears – seems like the least bad option.

The polar bear case is particularly stark due to the combination of the bears' charisma and the threat they pose to human communities. In other kinds of cases, the animals in need of food could simply be ignored. To do so will, in many cases, be the easy or default option. But to take this option is to ignore the role that we may have played in the fact that these animals need food. It is one thing to decide to ignore animals in need when they are nothing to do with us – but it is quite another to ignore animals when they are in need and it is our fault. Equally, perhaps it is legitimate (if callous) to pass by a hungry human being in need and offer no support. But it is deeply wrong to refuse to help a hungry human after *we* have knocked their last food from their hand, or poisoned the vegetable patch on which they depend, or ruined their stores by leaving a door open.

But perhaps there is a disanalogy between the case of knocking food from a stranger's hand and – through our consumption practices – causing a drought that leads to wild animals starving. In one case, there is a particular human who is responsible. There is a very straightforward story about moral and causal responsibility, and a very straightforward conclusion about who is responsible for rectifying the situation – if I knock your sandwich from your hand, it is my responsibility to remedy the situation. But in the other case, stories about causal and moral responsibility are much more complicated (Palmer 2020, 452). The situation mirrors some of those discussed in chapter 4, but is more complex still. There is already a well-developed debate about who – in practice, which state – is obliged to pay for the impacts of climate change on *humans*, and there is no reason to think that questions in the animal case should be all that different. The point is that, when animals are unable to feed because of the catastrophic impact of anthropogenic climate change, they have a compelling case that their rights have been violated – and there seems to be *some* human (group) who is morally responsible for feeding them. We (collectively) have entered into a morally salient relationship with them by destroying the things that they need to survive.

Of course, this is not to say that we are *always* in such a relationship with wild animals in need. But when we are, we are obliged to help them. And when we *might* be, it could be the responsible thing to do. And even when we are not – and know we are not – we may still have the duty of beneficence explored in the previous section. So not only – assuming a duty of benevolence – is there always some reason to feed wild animals in need, we (collectively) will often have an obligation to do so.

Again, though, I stop short of fully endorsing the feeding of wild animals. First, doing so may be rights-violating. We can raise worries about feeding food acquired in rights-violating ways, as we explored throughout this book. And we can raise worries about fostering dependency, as explored in chapter 4. And we can raise worries about respecting the habitat rights of these wild animals, not exerting undue control over their spaces. And we can raise worries about making animals accustomed to humans, putting those animals (or other humans) at risk. This is just a taste of the wide variety of ethical dilemmas we will have to negotiate if we choose to feed wild animals. Without a clear indication that we have an *obligation* to feed (thus, in this case, an indication that we are collectively responsible for the plight faced by the animals), it might be best not to become involved, parallel to how staff of wildlife rehabilitation centres might reasonably choose not to become involved with particular animals, as explored in chapter 6.

Second, we need clearer ideas of the institutional mechanisms that could be put in place to allocate responsibility for feeding (and the shouldering of the costs of feeding), on top of the institutional proposals for feeding regimes that are not (or, recalling our 'catastrophic moral horror' proviso, are only minimally) rights-violating. And third, we are not able to say just how many wild animals are facing hunger because of our actions. In some cases – for example, animals who are real niche specialists whose niche is vanishing because of climate change, like the polar bear – it will be obvious. But in other cases, perhaps many or most cases of wild animal hunger, we are not responsible for this hunger, and have no normatively significant relationship with the animal in question, and so are not obliged to intervene.

Concluding Remarks

In this chapter, I have explored a surprising question, and reached (what may seem to be) a surprising conclusion. Wild animals are the animals most distant from us, and with whom we seemingly have the fewest morally salient relationships. On the face of it, then, we seem to have no obligations to intervene to protect wild animals from becoming each other's food, and to provide food for wild animals in need, even while we do have obligations not to interfere with them – and to stop our companion animals from interfering with them.

This is, in an important sense, right. But it is also, in two important senses, incomplete. We may well have imperfect duties of beneficence to aid wild animals. After all, death in the teeth of a predator or death from starvation involve a great deal of suffering. If it is within our power to remove this suffering, we might think it better, all else equal, that we do so. This is not the same as saying that we are *required* to act to protect animals from predation or starvation, but might be to say that it would be *good* if we did. I have not endorsed the claim that we do have this duty, but it is surely plausible that we do. And, despite appearances, we may often be morally and causally responsible for the suffering faced by wild animals. If we are, we are responsible for putting things right.

But I endorse a note of caution. When it comes to intervention – especially for beneficent reasons, but also to put right suffering for which we are responsible – we must be careful to respect the rights of wild animals. These might include rights to a degree of control over their own spaces, and will certainly involve rights against violent interference. A great deal of further work and thought is required before wholescale intervention can be endorsed. But, perhaps, this is work that we should be doing. And given the theoretical nature of this chapter, I have not spent much time reflecting on the actual interventions that people *do* offer to aid wild animals in need. Many of these, to be clear, are surely praiseworthy. Some may not be. The situation is – perhaps dissatisfyingly – complicated.

It is worth remembering, too, that this analysis of our relationship with wild animals has a bearing on many other animals. If we have imperfect duties of beneficence towards wild animals, we presumably have them towards the animal neighbours of chapters 4 and 5, too. And it is not just free-living animals who are impacted by climate change. Any number of animals, from those closest to us to those furthest away, may find themselves going hungry because of what we (collectively) have done to the planet. The considerations of this chapter, then, have the potential to impact the conclusions of each of those that has come before. The challenge of wild animal suffering has the potential to upturn many settled assumptions about animal rights.

8

Conclusion

Throughout this book, I have addressed the ethics of feeding animals by exploring a range of relationships that (particular) humans can and do have with (particular) animals. This exploration has elucidated and defended a novel normative position. According to this position, animals have negative rights on account of their possession of morally significant interests. For example, all sentient animals have an important interest in not being made to suffer, and typically possess an important (though greater or lesser) interest in continued life. They thus have rights against being made to suffer and against being killed. Consequently, we wrong animals by making them suffer or by killing them. Not only do we wrong them, but they are victims of an injustice.

These rights and wrongs can do lots of important work when it comes to thinking about food obligations concerning the eating of animals. And these are the kinds of questions that are usually addressed by food ethicists thinking about animals and animal ethicists thinking about food. But they are not much good for thinking about the *feeding* of animals. If we think only about the interests of animals when it comes to *positive* rights and duties, we potentially arrive at alarming conclusions about how – in principle – we have as much obligation to feed wild animals as companion animals. Instead, I have suggested, we need to think seriously about the different kinds of relationships that we have with animals to understand their positive entitlements, and our positive duties towards them. These relationships can take many forms. Many of them are causal: in what sense are we causally responsible for vulnerability or need experienced by the animal? In what sense are we morally responsible for this vulnerability or need? Many others will involve affect. In what sense have we welcomed animals; in what sense are they our friends? Yet others will involve questions of membership and belonging. Are these animals in some sense part of our group or community? Is our group or community in some way responsible for the needs experienced by this animal, or the harms they might cause?

I do not pretend to have explored every possible kind of relationship that could be morally salient when it comes to determining the positive entitlements of animals. And nor do I pretend to have explored every situation in which the potential (non-)feeding of animals raises ethical questions. But I do hope that I have provided some important answers, and the building blocks needed to reach others. For example – and to return to a point made in the introduction – I have said nothing about the obligations owed to farmed animals. And I have not said anything about the obligations owed to animal workers. But by thinking through the kinds of conclusions reached concerning (for example) companion animals on the one hand and animals in wildlife rehabilitation centres on the other – and by combining this with consideration of the particular features of the relationship we or others have with farmed animals or animal workers – we can reach, I believe, clear conclusions about the feeding of these animals.

In this book's closing remarks, I think it worthwhile to highlight some of the conclusions that we should take away from the preceding analysis – analysis that, at times, may have felt abstract. Some of these conclusions, considered without the preceding analysis, may seem mundane. Some may seem alarming. Some of them focus on our own individual behaviour, and some of them think of us as socio-political actors, able to influence various institutions.

1. We should look seriously at switching our companion animals to a plant-based diet (if they do not already have such a diet). For some animals, this will be easier than for others. If we are not confident that we can switch our companions to a plant-based diet for whatever reason, we should seriously explore assorted 'alternative' diets for our companions. In years to come, it will perhaps be easy to feed our companions a diet

based upon the products of cellular agriculture or animals we are certain are non-sentient. Here and now, we can at least explore the possibility of invertebrate-based foods, or of sourcing animal-based foods in ways that will not put money into the pockets of animal agriculturalists.

2 We should support the scientific (but not just scientific) exploration and development – in animal-friendly ways – of cellular agriculture, plant-based diets for carnivorous animals, and the farming of non-sentient animals (including exploration of *which* animals are non-sentient). Perhaps we do not want to support these things for reasons of mere scientific curiosity, or in the interests of diversifying and expanding human diets, or to develop new revenue streams. But we *should* want to explore these things for the potential that they have to create safe and animal-friendly means to feed carnivorous animals.

3 We should – thinking here of us in our capacities as animal activists and members of animal activist organizations, if we are – be more willing to talk about animal-friendly animal foods. The thought of plant-based diets for companion animals is not as alien as it used to be, but can still evoke incredulity or wrath. If we are armed with the facts, there is no reason that these topics should not be broached. Once large animal organizations start to address these issues, on-the-ground animal activists and vegans will begin to gain a better understanding. And, from there, these topics will be normalized.

4 We are obliged to feed our companion animals. This is not surprising. But we should be clear about which values should inform our feeding of companion animals. We should not be worried about what is putatively natural for our companions. But – and though we must again stress the need for foods that are respectful, both to animals and to humans – we should be open to giving our companions a degree of choice over what they eat. At the same time, we must be prepared to limit what our companions eat. We wrong our companions if we make them ('allow them to become') obese. This is, realistically, a greater threat to many companions than going unfed.

5 We have a duty to prevent our companions from *becoming* food, but so too we have a duty to prevent our companions

from making other animals *into* food. We must take steps to ensure that our companions do not (successfully) hunt. If our companion dogs kill rabbits or our companion cats kill birds, rabbit and bird blood are on our hands.

6 We are permitted to welcome our animal friends (for example, garden birds) into our spaces by feeding them. We are not required to do so, but if we *choose* to do so, we take on certain (minimal) responsibilities for these friends. Again, we must limit what we feed them, out of a need to respect both humans and animals in the sourcing of food. And we must be careful not to put animals at undue risk, to foster dependency, or to overstep our marks as hosts. Crucially, too, we are permitted to *exclude* animals who we do not want in our space, providing we do so in a way that is respectful. Keeping rodents out of our stores using secure doors or discouraging them using smells they find unpleasant is one thing. Deadly traps and poisons are quite another.

7 We should not lose sight of the fact that even plant-based foods are harmful to animals. Farmland, even arable farmland, can be an unsafe place for animals. Beyond growing our own food (where possible), there is not much that we as individuals can do. But a gradual transition towards indoor agriculture could save many animal lives. Again, it is important for us to be *aware* of these harms and for animal organizations to address these harms. Not only could greater awareness lead to solutions, but greater awareness will disarm those critics of veganism who disingenuously (or even sincerely) present plant-based diets as more harmful to animals than meat-based diets.

8 We need to push back against the feeding of harmful foods to animals – especially the feeding of animal-based foods to animals – in *all* of the various institutions in which animals are kept. This includes wild animals who are rescued from situations in which they are in dire need, whether anthropogenic or otherwise.

9 We need to move away from ideas of releasing 'rehabilitated' predatory animals. When we rehabilitate predatory animals, we help them to hunt and kill successfully. This makes us partially

responsible for the harms that they go on to commit. While it may be admirable to rehabilitate wild predators in need, re-releasing these animals cannot be the ultimate goal.

10 And, while I am far from saying that we need to reshape nature, protect prey from wild predators, or feed every animal in need, we must acknowledge that the conversation about wild animal suffering is one that we should be having. If we begin research in earnest now – and if this includes philosophical and social scientific research about the ethics of intervention in nature – we will find ourselves with a clearer understanding of how (if at all) we could and should intervene to provide wild animals with food, and protect them from becoming food, in the future.

Creating a fairer and more respectful world for animals requires changing our food practices. That much has been clear for many years, and has not been my focus in this book, or in these recommendations. What has been less clear is that we also need to change our *feeding* habits. Just as our diets need to change for the world to be a fairer and more respectful place, so might the diets of animals. If we were to take these ten messages to heart, we would do a great deal to move our world, presently a dark place for animals, to something a little brighter.

Notes

Chapter One

1 For readability's sake, I will follow the convention of using *animal* to refer specifically to *nonhuman animals*. Humans are animals, too. If I use *animal* more broadly, it will be contextually evident.

2 The thirteenth-century Italian priest St Thomas Aquinas, the most influential figure in the tradition of scholastic philosophy, claims to 'refute the error of those who claim that it is a sin for man to kill brute animals. For animals are ordered to man's use in the natural course of things, according to divine providence. Consequently, man uses them without any injustice, either by killing them or by employing them in any other way' (quoted in Fischer 2019a, 1790–1). The seventeenth-century English philosopher Thomas Hobbes gave animals short shrift in *Leviathan*, his *magnum opus*, noting simply that 'to make Covenants with bruit Beasts, is impossible; because not understanding our speech, they understand not, nor accept of any translation of Right; nor can translate any Right to another: and without mutuall acceptation, there is no Covenant' (1996, part 1, chap. 14, para. 22). Given that Hobbes's ethics and politics rest upon covenants (contracts), this indictment is damning indeed. The eighteenth-century German philosopher Immanuel Kant – in an intriguing piece entitled 'Duties towards Animals and Spirits' – similarly characterizes humans as owing nothing to animals directly: 'all animals exist only as means, and not for their own sakes, in that they have no self-consciousness, whereas man is the end … it follows that we have no immediate duties to animals; our duties towards them are indirect duties to humanity' (1997, 212). Kant can be praised for recognizing that cruelty and wanton destruction of animals is wrong, but should be criticized for refusing to recognize that we owe something *to the animals themselves*.

3 Famously, the eighteenth-century English utilitarian Jeremy Bentham speculated that 'the day *may* come, when the rest of the animal creation may acquire those rights which never could have been withholden from them but by the hand of tyranny'. When it comes to the question of obligations to others, Bentham recognized, 'the question is not, Can they *reason?* nor, Can they *talk?* but, Can they *suffer?*' (1996, 310n1, emphasis in the original). These claims were later defended by the philosopher and social reformer John Stuart Mill (1867). A passage similar to Bentham's can be found in the preface of the Genevan theorist Jean-Jacques Rousseau's *Discourse on the Origin and Foundation of Inequality among Mankind*, in which he writes that 'as [animals] partake in some measure of our nature in virtue of that sensibility with which they are endowed, we may well imagine they ought likewise to partake of the benefit of the natural law, and that man owes them a certain kind of duty. In fact, it seems that, if I am obliged not to injure any being like myself, it is not so much because he is a reasonable being, as because he is a sensible being; and this quality, by being common to men and beasts, ought to exempt the latter from any unnecessary injuries the former might be able to do them' (2002, 85).
4 Exceptions include Henry Salt (1980), an English philosopher and vegetarian who published *Animals' Rights* in 1892, and Porphyry (2000), a third-century Neoplatonist philosopher, who authored *On Abstinence from Killing Animals*.
5 A term he borrowed from Richard Ryder, another of the Oxford Vegetarians.
6 Though Singer is a utilitarian, there is an open question about the extent to which *Animal Liberation* is, strictly speaking, a work of utilitarian philosophy. Singer writes that 'the text of *Animal Liberation* is not utilitarian. It was specifically intended to appeal to readers who were concerned about equality, or justice, or fairness, irrespective of the precise nature of their commitment' (Singer 1999, 283). This is disputed by Garner and Okuleye (2020, 117–19), who argue that *Animal Liberation* is indeed utilitarian.
7 Often, the Singer/Regan approach is called an 'interest-based' approach, but I shall not mirror this terminology, for reasons that will become clear shortly.
8 For examples and extrapolations of these ideas, see the papers appearing in the collections edited by Carol Adams and Josephine Donovan published respectively in 1995, 1996, and 2007.
9 Characterizing Regan as a partially political thinker may seem surprising, but see Milburn 2016a and Cochrane, Garner, and O'Sullivan 2018.

10   To be clear, interest-based rights approaches to animal ethics are not completely new – they (or something like them) have received defence (Feinberg 1980, chaps. 8–9; Rachels 1990; Sapontzis 1987) and discussion (e.g., Regan 1976; Frey 1977) in twentieth-century animal ethics. However, as a key mainstream framework, they represent a new, promising, and compelling development.
11   Of course, when it comes to both humans and animals, there is room for argument about who gets into what category, precisely what rights and responsibilities each category entails, and so forth.
12   Indeed, the way the philosophical examination of food is framed frequently implicitly excludes questions about animal diet, furthering my worry that the literature is importantly incomplete. Paul B. Thompson suggests that food ethics (which he understands broadly to include topics beyond moral philosophy) 'encompasses debates over the production, consumption and cultural significance of the *human* diet and the technological apparatus that supports it' (2016, 61, emphasis added). Similarly, alluding to a discussion between Glaucon and Socrates in Plato's *Republic*, Fritz Allhoff and Dave Monroe neatly separate issues of human diet and non-human diet, suggesting, it seems, that animal diets are of no philosophical interest. 'To give food a just, properly nuanced, philosophical treatment', they say, 'requires sustained investigation: we are, as Glaucon indirectly observes, more than mere pigs, so discussion of our diet calls for more sophistication' (2007, 2). This dismissal, however, is not universal (see, e.g., Ceva and Bonotti 2015, 400–1; cf. Milburn 2015a; 2016a; 2019a), and philosophers of food have often left their accounts of what is a part of their field open-ended. So, for example, in his introduction to *The Philosophy of Food*, David Kaplan does not mention animal diets, but says that 'there is much more to the philosophy of food' than those areas he outlines, noting that he 'probably left out more than [he] included' (2012, 18). Meanwhile, Matthias Kaiser and Anne Algers characterize food ethics 'as a dynamic field with ever new topics and challenges arising' (2016, 6). The lack of attention paid to animal diet in the literature in the philosophy of food is surprising, as the topic has not been overlooked in food studies, where the industries around animal food are recognized as part of the food industry more generally, and so worthy of both attention and critique. For example, Marion Nestle, a leading scholar of and advocate for food studies, authored *Pet Food Politics* (2008), exploring the pet food industry analogously to (and as a part of) her examination of human food politics, and later co-authored *Feed Your Pet Right* (Nestle and Neisheim 2010).
13   Some of his explorations were decidedly ethical, including his consideration of gluttony; people eating too much, however, is also – for

Plato – metaphysical and epistemological, insofar as it raises questions concerning the relationship of the 'appetitive' part of the soul to other parts. Plato also asked political questions, interrogating the diet in the ideal city-state. Interestingly, the diet of one envisaged (commended?) state is entirely vegetarian, though not vegan (see Dolgert 2018; Dombrowski 1984; Painter 2013; Silvermintz 2019).

14 A political philosophy of food can be separated from the *food justice* literature, which asks (primarily) about disparities in access to or control over food along gendered and racial lines. It does this by raising questions, for instance, about whether certain groups are excluded from food distribution (at the international, national, local, or even household level), whether food producers/workers are offered decent conditions and fair remuneration, and whether the burdens of food production are equitably shared (Gottlieb and Joshi 2010; Dieterle 2015). Perhaps due to the traditional hostility towards animal advocacy among other justice movements (Donaldson and Kymlicka 2014), vegans and animal advocates have not always been welcomed into the food justice movement/literature. Despite this, there is nothing to prevent the greater incorporation of veganism, animal rights, and animal ethics into the food justice movement *in the future*. I say this because 'exactly what constitutes a food justice approach still remains a moving target' (Gottlieb and Joshi 2010, 4–7). In addition, it is undeniable that, first, animal advocacy/the vegan movement has a lot to learn from other justice movements, including the food justice movement, as it critically reflects on the injustices to *humans* that it perpetuates and endorses (see Harper 2010; Wrenn 2016), and, second, that one can engage in food ethics – explicitly understood as *separate* from food justice – with a full consciousness of the importance of intersectional activism and scholarship (Thompson 2016).

15 For a contrary view, see Ahlhaus and Niesen 2015.

16 This is consistent with what I have said about Plato's discussion of diet in the ideal state; Plato is mentioning food as a part of something else/ in his pursuit of some other goal, rather than considering food in its own right (Allhoff and Monroe 2007, 1–2).

17 Some philosophers examining food go further still; Raymond D. Boisvert and Lisa Heldke, in an ambitious move that almost reverses the assumed relationship of discipline and object of study, aim to rethink philosophy around food and our status as eating, stomach-endowed beings (Boisvert 2014b; Boisvert and Heldke 2016).

18 As I will hint in the second chapter, there may be room for something *like* animal agriculture in a future state. But the details of this are yet to be filled in. To spend time exploring how we might feed these

hypothetical 'working' animals without a clear idea of what this relationship would look like would be to put the (hay-filled) cart before the (working) horse.

Chapter Two

1. Though I will use the less derogatory *companion* to refer to companions themselves, I will use *pet food* to refer to pre-made products intended for the consumption of companions, and *pet food industry* to refer to the industry that makes these products.
2. This dichotomy is muddied by animals like rats, who are simultaneously thought of as companions and pet food (Pierce 2016, 80).
3. Many of these works present arguments for 'vegetarianism', but the thrust of the argument seems to support veganism rather than vegetarianism. Indeed, it is difficult to come up with an argument for vegetarianism that does not function more consistently as an argument for veganism (cf. Milburn 2019c), given the death and suffering inherent in animal agriculture generally.
4. Potential technological fixes – such as sexing chickens' foetuses while they are still in the egg – provide some promise of respite.
5. For more on this, see my discussion – and the references – in Milburn and Van Goozen 2021.
6. Carnism is the culturally variegated ideology that constructs some animals as edible, and some animals as inedible; some animals as food, and some animals as something else. There is nothing necessary about the fact that certain Westerners eat cows but not dogs; some Koreans will eat dogs, while certain Indian groups will not eat cows, even though they may eat other animals (Gibert and Desaulniers 2019). Veganism and carnism *qua* ideologies do not necessarily neatly map onto veganism and non-veganism *qua* practices; for example, a carnist might take up plant-based eating for health reasons, meaning that they are carnist ideologically but vegan (or, perhaps more precisely, *plant-based*) in diet. Certain humans will lack an ideological commitment to carnism or veganism; infants, for example, are simply fed, and have no ideologies about food (though may prefer certain foods). For many carnists, however, their commitment to carnism will be present, but hidden.
7. None of this is *necessarily* so.
8. Variants are offered, for instance, by Jan Narveson (1999, 140), and Loren Lomasky (2013, 177–200) – though whether these particular philosophers *do* believe that we need a good reason to inflict suffering upon animals is perhaps not clear.

9   Outside my own work on the topic (2015a; 2017d; 2019a), worries have been raised, for example, by Donaldson and Kymlicka (2011, 149–53), Jan Deckers (2016, 98–9), and Jessica Pierce (2016, chap. 18). Jed Gillen, who runs a business selling plant-based pet food and associated products, majored in philosophy before dropping out of college. He peppers his very readable *Obligate Carnivore* (2003) with mentions of Socrates and Plato, as well as some animal ethics.

10 Going into the details of this point will take us too far from the present enquiry. Compare the discussion and references in Milburn and Van Goozen 2021.

11 An alternative means of (putatively) justly solving the problem of carnivory by removing those carnivores to whom we have an obligation comes in the form of genetic engineering; we could develop ways of engineering organisms away from carnivory altogether (cf. Johannsen 2017; McMahan 2010).

12 I have encountered people who are hostile to the idea that dogs are omnivores, but, biologically speaking, this is the correct classification; domestic dogs, compared to (more) carnivorous wolves, have co-evolved with humans by developing omnivorous feeding habits (Nestle and Nesheim 2010, chap. 2).

13 Conversely, it is conceivable that some omnivores cannot survive without animal protein.

14 It is worth acknowledging that there are a tangle of legal questions that might arise through this kind of scavenging. For example – to use an extreme example – the law is unlikely to look kindly upon a well-meaning animal activist 'scavenging' human flesh so that it can be fed to wild animals in need.

15 Again, accepting the permissibility of animal consumption of human bodies does not commit us to saying that *humans* may consume human bodies. The ethics of cannibalism are interesting, but beyond the scope of the current enquiry. Compare the discussion of cultivated meat below.

16 This is how Ed Winters ('Earthling Ed'), founder of the UK-based Surge Activism, summarizes his case against eating the eggs of backyard chickens: 'ultimately, the most ethical way to look after backyard hens is a way that provides them the best opportunity to live a long, healthy and happy life, and by taking their eggs we compromise that, meaning that taking their eggs is less ethical than feeding them back, or taking safe steps to reduce the number of eggs they produce in the first place.' Why feed the eggs back to the hens? 'One of the best ways to protect hens from nutrient deficiencies and health problems', Winters explains, 'is to

feed their eggs back to them, which will help the hens replace their lost nutrients, and they also really enjoy it? See https://www.surgeactivism.org/backyardeggs. Now, good hen care will of course include ensuring that the hens are taking on nutrients lost to egg production. One way to help with this is feeding eggs back to chickens. But this is not necessary to achieve the goal. Good hen care will of course include ensuring that they have food that they enjoy. One way of doing this is feeding eggs back to chickens. But, again, this is not necessary to achieve the goal.

17 This may sound unsatisfying, but we must remember that the world as we find it today is already some way from where it needs to be; think of how few people are vegans in even the most animal-friendly countries.

18 Responses to objections that have been grounded in other ways are discussed by others (e.g., Hopkins and Dacey 2008; Schaefer and Savulescu 2014).

19 Though that's not to say that any of these objections are decisive. Again, a stance on cannibalism is beyond the scope of this chapter, but cf. Milburn 2016b.

20 As of December 2020, cultivated meat is available for sale in a Singaporean restaurant, as well as available (though not for sale) in an Israeli restaurant.

Chapter Three

1 Some readers may initially think this trivial; those in the urban/suburban West may not think that their dogs and cats are in any particular danger of being predated. Of course, our dogs and cats do risk predation in certain environments; bears, wolves, large felids, and even birds of prey may predate upon cats and dogs. Of more particular concern, however, are smaller companions. Despite spending my childhood in rural England, hardly known for its ferocious predators, I recall neighbours worrying about companion rabbits being predated by foxes and companion fish being predated by herons.

2 Surprisingly, this view is controversial. Gary Varner (1998) at one time held that plants have interests based on claims about biological functionality, while Chris Belshaw (2016) extends this idea to non-living artefacts. I reject these notions. I consider the teleological claims upon which they rely problematic, and it seems clear that nothing can be bad *for* an entity that does not have conscious experiences. Sentience is sufficient for conscious experiences, and may also be necessary. In this sense, I hold that an organism's possession of interests requires that there is, to borrow a phrase from Thomas Nagel, 'something that it is like to be

that organism – something it is like *for* the organism' (1974, 436). There is nothing that it is like to be a chair or a plant (Lamey 2019, chap. 9).

3 This description of what might make certain feeding practices undignified draws upon Lori Gruen's account of wild dignity, according to which we 'dignify the wildness of other animals when we respect their behaviors as meaningful to them and recognize that their lives are theirs to live' (2011, 155; cf. Gruen 2014). Gruen explicitly does not apply this account to companion animals; they are not wild, so cannot have wild dignity. Nonetheless, it captures something about a common view of the ethics of feeding animals.

4 In context, she talks of *human* dignity, but there is no reason that dog, rabbit, or goldfish dignity should be different.

5 There is a split in the animal ethics literature (Crary 2018). Mainstream animal ethics accounts, including paradigmatic utilitarian and animal rights approaches, are *morally individualist* (Rachels 1990), and thus derive claims about the obligations we have to a given being from consideration of psychological, physiological, or relational facts about that being. This is challenged by non-individualist accounts of animal ethics, such as the Aristotle-inspired accounts of Elizabeth Anderson (2005) and Martha Nussbaum (2006), or the Wittgensteinian accounts of Cora Diamond (1978) or Alice Crary (2016). These accounts instead (or additionally) look to facts about the kind or group to which the individual belongs, with species membership standardly being of considerable import. Traditionally, moral individualists have been hostile to talk of dignity. So, Singer says that 'philosophers frequently introduce ideas of dignity … at the point at which other reasons appear to be lacking, but this is hardly good enough. Fine phrases are the last resource of those who have run out of arguments' (1974, 113), while James Rachels (who popularized the term 'moral individualism') argues that 'Darwinism leads inevitably to the abandonment of the idea of human dignity and the substitution of a different sort of ethic' (1990, 171). On the other hand, the critics of moral individualism use the language of dignity quite freely; indeed, dignity is sometimes presented as something that moral individualists have missed.

6 It could be the case that animals have an indirect interest in dignified treatment, even if they have no interest in dignified treatment in and of itself, because undignified treatment is likely to lead to a diminishing of their status in the eyes of others, and perhaps – in turn – harmful treatment. As Suzanne Laba Cataldi says of dancing bears (the paradigmatic example of animals treated in an undignified way), it 'is literally laughable – with the bears dressed up and behaving in such a

"silly" manner – to think of them as our "equals" and entitled to the same respect. The whole idea of animal rights becomes amusing in this context, is made to seem ridiculous. (Re)assured by their "own" performance of their "inferiority", their unequal treatment appears to be justified' (2002, 118).

7. It was developed by humans around 11,000 years ago, while humans have existed for around 300,000 years. Throughout human history, there have been human communities of hunter-gatherers who have lived without agriculture, including today.

8. These obligations are recognized in the food literature. Unfortunately, they are sometimes framed in terms of locavorism and 'food miles'; problematic notions that do not have the environmental significance and merit that their proponents seem to assume (McWilliams 2009, chap. 1; Goodnick 2015; Ferguson and Thompson 2020).

9. For more on the environmental impact of companion animal diets, see Sandøe, Corr, and Palmer 2016, chap. 14, and Oven, Ward, and Bethencourt 2020, chap. 2.

10. As in human obesity, certain health and/or genetic factors can predispose an individual to obesity, and, as in the human case, there are complex sociological factors that can increase an individual's risk of obesity (see Sandøe, Corr, and Palmer 2016, chap. 8). None of these things should lead us to overlook the large share of responsibility that guardians have concerning their companions' obesity.

11. This is not to say, incidentally, that there is anything *necessarily* wrong with so-called *ad libitum*/free feeding. For example, ensuring that a hamster always has access to food and especially water is surely appropriate.

12. It is now well recognized in the food literature that parents who feed their children poor-quality diets may not be blameworthy for this fact, and my comments about parents who fail to feed appropriate diets to their children are not meant to endorse the kind of naïve moralizing that characterizes certain narratives in the popular dialogue on food.

13. Perhaps, analogous to questions about dignity in feeding, there are ethical questions to ask about the *reason* that companions are given limited choice in their diets. There are ethically defensible reasons for giving dependent family members – including companions and children – no choice over their diets. For example, the family may subscribe to a puritan ethico-cultural norm that limits the choice of food available; the family may have limited food choices due to economic circumstances; those responsible for providing food may be confident that limiting choice is the best means to ensure healthful diets; or the

family may conceive of food as simply fuel, giving its members ample choice and opportunity for enjoyment in other areas of their lives.
14. In *My Dog Always Eats First* – as suggested by the title – Irvine (2013) documents how homeless guardians in America reconceive companion/guardian relationships around a new set of principles, including the idea that the companions eat, even if that means the guardian going hungry. Indeed, Irvine documents how individuals and organizations are often very happy to help the homeless feed their companions, to the level that some homeless people rarely worry about acquiring pet food, though they may struggle to feed themselves.
15. For a discussion of the ambiguity in ideas of 'wildness', see Palmer 2010, 64–5.
16. Note that this is not the same as saying that the state is obliged to feed companions. Compare: The state can rule that cars must have seatbelts without committing to *providing* cars with seatbelts.

Chapter Four

1. Ecologists might dispute this use of the term. Insofar as we receive benefits from the presence of these animals, we have a *mutualistic* relationship with them. Insofar as we are harmed by their presence, they are *parasitic* – at least as ecologists generally use these terms. Of course, the senses in which we benefit from or are harmed by these animals may not match neatly with the sense in which ecologists would typically use these terms. Nothing much hinges on this at present.
2. Leaving aside, for a second, cultivated meat.
3. As discussed in the previous chapter, I am more inclined to think of companions as family members than as friends, but I allow that it might be appropriate to describe certain animals as friends in this richer sense – a good example is provided by the relationships that we have with the *companions of others* (a friend's dog, for instance, might be my friend in her own right).
4. Raymond D. Boisvert (2014b, chap. 3) attempts to reclaim the term *parasite* to refer to one who eats with someone else. Boisvert's term thus has promise as a term for a relationship grounded upon feeding, and, indeed, I think it would be fair to call the animal neighbours we feed *parasites* in Boisvert's sense. However, Boisvert's conception of parasitism is very far removed from our vernacular use of the term, which carries highly negative connotations. Indeed, both the vernacular and ecological sense of *parasite* better captures the animal *foes* whom we aim to stop eating the things we leave in their reach.

5 Might they be dependent upon us collectively? Perhaps, given that they have adapted to live alongside us. But we are not responsible for those adaptations in the way we are responsible for the adaptations of companion animals. Or at least not to the same degree. When the lines blur, difficult questions arise – as will be discussed later in this chapter.
6 Perhaps these problems can be resolved; my point is only that they need to be resolved by anyone who wants to claim that we have a moral duty to feed our animal neighbours.
7 Telfer explicitly links optional virtues to the 'old notion of imperfect duties' (1996, 96). Imperfect duties are duties that we have, but that we do not always have to exercise. To use a straightforward example, perhaps I have a perfect duty not to kill my houseguests, but only an imperfect duty to donate to their charitable causes.
8 Raymond Boisvert accepts that Telfer's characterization is the correct one for a *modern* philosophy, which he contrasts not only with a premodern philosophy in which hospitality has an important role, but also a postmodern philosophy in which hospitality *is* ethics, drawing upon the ideas of the French philosopher Jacques Derrida (Boisvert 2014a, 1187–8; Boisvert 2014b, 26). The present work – despite its atypical focus on relationships – is decidedly modern in Boisvert's sense.
9 This is even despite Boisvert's contention that Telfer's 'modern' approach to hospitality is less able to accommodate ideas about human relationships with animals (2014a, 1188). Boisvert's counter-suggestions, which conceive animals as a part of the natural 'other', seem to fall under the banner of what I would characterize as an 'environmental' approach to animals. While he (along with postmodernists and environmental ethicists) might seek ways to rethink humans and philosophy relative to some holistic 'other', I (along with other so-called extentionist animal ethicists) seek to explore the possibility of extending existing ethico-political concepts and tools to encompass animals as well as humans.
10 Extending hospitableness in all its senses to all and sundry is impossible – different exercises of hospitableness may be incompatible with one another (Telfer 1996, 99–100). More on this in the animal case shortly.
11 Drawing a clear line is difficult.
12 Note, too, that actually existing legal structures are not kind to animals perceived as 'wild'. Richard Posner (2005), for example, draws attention to a legal case in which Mia, a formerly 'wild' raccoon, was killed by authorities after being seized from her human guardian.
13 The exception comes about when I extend Good Samaritan hospitableness to my non-friend neighbours; in that time, I do acquire a particular moral duty of protection – the degree of which will vary.

14 It is also, if you like, a claim made in moral philosophy's voice, rather than one made in political philosophy's voice.
15 Peter P. Marra and Chris Santella (2016) open their *Cat Wars* with an 'obituary' of the Stephens Island wren, a species supposedly made extinct by the actions of Tibbles, a single cat. This story is typical of the problematic way that cat-on-bird violence is talked about. The obituary is for a charismatic *species* eliminated by Tibbles, not for the sensitive *individuals* killed by Tibbles. Tibbles has a name, a family, and a personality. The wrens exist solely as representatives of a species. On the picture I here present, it is individual animals – including animals from unloved species – who are the victims of injustices, and not species or other collectives.
16 Or, alternatively, that *both* are citizens; but given that Donaldson and Kymlicka analogize denizen animals to migrant workers, this does not seem to be their proposal.
17 Let us not worry too much about what I am feeding the sparrowhawks. Lab-grown sparrow flesh, perhaps. Soy-based raptor treats.

Chapter Five

1 I borrow the term from Andy Lamey (2007; 2019). A burger vegan is someone who accepts the ethical underpinning of veganism, but who nonetheless supports eating (beef) burgers.
2 It is worth noting that the idea that many of these crops are being grown to feed other animals is not something that is true only in a vegan utopia; indeed, a large proportion of certain crops – including, for example, soy and corn in a North American context – are grown and subsidized so that they might be fed to farmed animals.
3 Let us leave aside the possibility of voluntary human extinction.
4 Davis may be misinterpreting Regan. Angus Taylor argues that 'Davis badly misreads Regan', as the latter raises the 'minimize harm principle' only to 'immediately to reject it' (Taylor 2019). We will return to possible philosophical errors in Davis's reasoning below.
5 How many burger vegans *actually* accept animal rights? Some may; some others may be simply exploring unforeseen consequences of ethical frameworks to which they do not subscribe. Though interesting, this question has no bearing on the strength of the arguments.
6 The typical 'humane omnivore' (see Pollan 2006; cf. Stănescu 2016), if living up to her own ideals, consumes only a relatively a small amount of local, sustainable non-vegan food. This may not score highly on burger vegan grounds. The 'humane omnivore' may eat a moderate amount of

chicken, for example. A comparatively high number of chickens need to die for a meal of free-range chicken bodies in contrast to the number of cattle who need to be killed to feed a community on pasture-raised cattle bodies. This is simply because cattle are a lot bigger than chickens. The consistent burger vegan will presumably be more troubled by the humane omnivore's consumption of chicken and pork than the vegan's abstention from beef – especially if the chickens and pigs are fed the products of arable agriculture.

7 It is worth noting that similar arguments can be presented in the language of interest-based rights – they are not merely a quirk of Regan's system (compare, e.g., Cochrane 2012, 65–71).

8 Indeed, it might seem puzzling to claim that both arable agriculture and pastoral agriculture involve *wrongful* setting back of interests. This would seem to entail (assuming, for now, no third option) that there is nothing we could do without wronging someone. It would be, consequently, impossible not to act wrongly. It is one thing to say that any of our options involve *harms* – that seems correct. It is quite another to say that any of our options involve *wrongs*, which is a bitter pill to swallow.

9 If they can produce more food per plant, then the effect will be that less land is required, potentially putting fewer animals at risk. If they merely grow quicker, then perhaps the effect on the number of animals killed will be negligible; though quick-growing crops open the door to less space being used, it presumably means that there will be more harvests (and other interventions) on the land that *is* used. It is these harvests and interventions that lead to harm to field animals.

10 I thank Sue Donaldson for drawing my attention to this example.

11 Though the RSPB has its roots in the anti-fur movement of nineteenth-century Britain, it is today mostly a conservation organization.

12 Some burger vegans (e.g., Schedler 2005) explicitly call for people to grow their own crops where possible.

13 Glasshouses are not always particularly bird-friendly; the risk of bird collisions needs to be a part of the conversation about glasshouse design in a way that it is, presently, not.

14 Despommier notes their suitability, in contrast to 'too big and too clumsy' cattle and sheep, who 'require a different ecological setting' (Platt 2007, 83).

15 Worms can also be utilized in aquaculture, either as combined vermiculture/aquaculture (Kotzen et al. 2019, 325), or else as food for the fish (Robaina et al. 2019).

16 It is tempting to imagine that millennia of farming have allowed these animals to evolve alongside human farming techniques. But, realistically,

the intensive farming methods favoured today are only a few human generations old – nowhere near enough time for animals to have evolved something like the internalized dependency discussed in chapter 3.

17 That is, their parents have had some success – they have reached sexual maturity and successfully birthed young.

18 Might there be obligations to aid the wild animals who come to live on a more standardly rewilded space? If so, then perhaps the obligation would be to rewild in a way that is particularly friendly to the animals there. For example, predator-free ecosystems could be favoured. Research on such possibilities is at a very early stage. But no one said that respecting animals' rights would be easy. Compare Cochrane 2018, chap. 5.

Chapter Six

1 Tristan Derham and Freya Mathews (2020) have offered valuable analysis of the category of animal refugees, though my use is independent of theirs.

2 Jozef Keulartz (2016, 819) criticizes me for holding that there are predator/prey relationships in which humans are not implicated in the 'Anthropocene'. I concede that certain humans or human groups may have a degree of causal responsibility for pretty much any harm to animals that occurs in the world – I doubt, though, that this means that there is injustice in all of these harms. A tiny share of causal responsibility need not entail any moral responsibility, as discussed in chapter 4.

3 Not all such charges could be described in these terms, though. It would be odd to talk of restoring healthy orphaned animals to health, for example – their plight is caused by neither illness nor injury.

4 In a sense, this isn't for me to say. We should not be quick to prejudge the kinds of alternative relationships that could develop in rights-respecting WRCS.

5 I say *surprisingly*, as she is open to supporting industrial animal agriculture, which requires significant harm to animals indefinitely. Cultivated meat, on the other hand, need not necessarily entail a reliance on animal agriculture or vivisection – her first two objections to the possibility. Meanwhile, as discussed in chapter 2, it would not necessarily 'symbolically [normalize] oppressive behaviour towards other animals' (Wrenn 2018, 165), especially if it was produced specifically to feed carnivorous animals, and especially if it could be used as an opportunity to break down human/animal species divides (see Milburn 2016b; Bovenkerk, Meijer, and Nijland 2020). Even if Wrenn is reluctant to support the production of cultivated meat for human consumption,

there is surely good reason to follow some of cultivated meat's critics, like Jan Deckers, who have been convinced of the value of cultivated meat as a source of food for animals.

6 Very expensive possibilities, such as hyper-humane milk production, hyper-humane egg farming, and the hyper-humane farming of animals for their corpses (for more, see Cochrane 2012, chap. 4), even if hypothetically just – on which I am here making no commitment – would hardly be suitable solutions to the present problem. An interesting philosophical possibility for sure, but not particularly practical for cash-strapped WRCs. A possible exception would be if only very tiny quantities of animal protein were required, perhaps because the animals concerned were very small, and/or their diets could be supplemented with plant-based foods, even if they cannot consist entirely of plant-based foods. But, for the most part, these kinds of approaches are not going to be suitable solutions to feeding the residents of WRCs.

7 In the last chapter, I spoke of mice being kept out of a compost bin. Might the releasing of predatory animals be a similar case? In both cases, animals are put at risk by the actions of humans. But I resist this comparison. Even if the *outcomes* may be similar, there is a big difference between putting someone at risk by denying them access to something to which they were never entitled and putting someone at risk by introducing (or in some way supporting/maintaining/saving) a major threat to them. Compare: even if my preventing A from taking some of my unused seeds (her usual approach) means that A cannot grow crops this year, my prevention is a very different affair, normatively speaking, than my salting or otherwise poisoning B's ground, preventing B from growing crops this year. Keeping mice out of a compost bin is like refusing to share seeds; releasing a cat among the literal pigeons is more like salting a neighbour's patch.

8 Ultimately, after the AI takes control of part of the ship and the crew engage in a range of activities aiming to disarm it, the crew manage to convince the AI of the fact that its orders were rescinded – which is true – and the AI self-destructs to destroy a series of *other* wayward bombs that have arrived. This is all rather neat, allowing the *Voyager* crew to circumvent the difficult moral question with which they are faced.

9 Although the Prime Directive, which necessitates Starfleet crews' non-intervention in the affairs of alien species in the *Star Trek* universe, seems to be ignored in more episodes than it is followed.

10 Indeed, for all the part that the AI's creators play in the episode, they may as well not exist.

11 The example is adapted from an earlier paper (Milburn 2015b, 286–7). There, I argued that the release would be legitimate. I no longer think this.
12 It is worth saying that Abbate's guardianship principle, if this is something we should accept, certainly makes a difference in the stoats-saved-from-hoarder case. The stoats are clearly the victims of injustice, so causing/allowing deaths to keep them alive may be justified. However, once again, this is presumably the case only if there is no other option. Given the possibilities that WRCs have for feeding carnivorous animals, discussed above, it seems likely that there are other ways to feed them. Given that releasing stoats would result in deaths for which humans are responsible for the benefit of the predatory animals, it seems like the Reganite framework that Abbate is developing will endorse my conclusions about not releasing rehabilitated predatory animals, even when they would be perfectly able to survive.
13 And this holds even if the would-be murderer and their would-be victim are foreigners, members of (as Regan says of wild animals) 'other nations' (2004, 357).

Chapter Seven

1 Perhaps concern with wild animals themselves may lead to the conservation of species or ecosystems for the sake of individual sentient animals who depend on them. On the other hand, concern with individual sentient animals could lead to the deliberate extinction of species (either for the sake of the members themselves, or for the sake of sentient animals the members threaten) or the upturning of ecosystems (as they often depend on massive levels of animal death and suffering for their functioning).
2 Both in the sense of the majority of species and the majority of individual animals.
3 There is thus a disanalogy between most wild animals and most distant human strangers. Distant humans are likely to be entangled with us in a whole variety of ways, through economic and social systems; through being parts of overlapping circles of personal relationships (my friend's friend's friend is the distant stranger's friend's friend's friend); or even simply through mutual awareness. Now, I do not take this to show that we always have a particular kind of relationship with other humans. Nor do I take it to show that we are always (morally, affectively, causally...) more closely related to other humans than to wild animals. But I do take it to *indicate* that we are frequently closer (in assorted morally

salient ways) to distant humans than to distant wild animals. There are undoubtedly many counterexamples.

4. To take an easy example, one could resist a duty of beneficence to animals while remaining non-speciesist by denying that we have a duty of beneficence to aid humans in need.

5. Or it can only do so when the pay-offs are suitably great. Gazelles have much less of an interest in continued life than do most humans; they lose a lot less when they die than we do (see Milburn and Van Goozen 2021). It would be disproportionate to demand that a human put themselves in mortal peril to save a gazelle to whom (or concerning whom) they have no special duties.

6. Does the lion have a right not to be scared off by a gunshot? Not to be glanced by a bullet to stop the chase? Not to be tranquilized? These are all interesting questions.

7. And, just as special obligations towards a dog could justify visiting violence on a human attacker, it is at least plausible that we could construct a scenario in which I am permitted (or even required!) to *save* a dog over a human.

8. For proposals concerning property rights, see Hadley 2016; Bradshaw 2020; Kianpour 2020; Milburn 2017a. For sovereignty rights, see Donaldson and Kymlicka 2011; Wadiwel 2015; Goodin, Pateman, and Pateman 1997. For more mixed accounts, see Cooke 2017; Milburn 2016c.

9. One might object that they cannot enjoy these spaces if they are torn apart by predators. This is true, but it does not mean that the interest grounds an obligation to protect them from predators. The distinction between negative and positive rights is again relevant. Your interest in being able to visit lots of interesting places might be (part of) what grounds your right to freely travel, meaning I am obliged not to stop you from travelling (without good reason). But that same interest does not obligate me to *help* you travel – give you lifts, pay for your train fare, etc.

10. Again, it could be objected that we would not normally use ideas of property rights to defend being torn apart by predators. This is true. But we *would* use ideas of property rights to object to people marching onto our land to build fences, even if they insisted (rightly or wrongly – it doesn't matter) that such fences would help protect us from predators.

11. This point is made in a variety of places. In one paper, Donaldson and Kymlicka write that their account 'insists that the moral purpose of sovereignty *is to prevent injustices*, in particular injustices of colonization, despoliation, or domination' (2015b, 340–1, emphasis in the original). In another, they stress that 'when humans conquer, colonize, settle and develop [wild animal] territories, they harm free-living animals not only

by killing them or reducing their food supply but also by denying them the right to maintain the ways of life they have developed in relation to their territory'; the recognition of sovereignty 'would serve as a powerful check on this injustice' (2013b, 215).

12 To be clear, it may be – it just happens that (currently, to the best of my knowledge) no one is feeding polar bears a plant-based, or otherwise rights-respecting, kibble.

13 Incidentally, bears are the kinds of wild animals who may well have particularly good lives, relatively speaking. It could be that bears will have an interest in going on living *even if* lots of wild animals will not.

References

Abbate, Cheryl. 2016. 'How to Help When It Hurts: The Problem of Assisting Victims of Injustice'. *Journal of Social Philosophy* 47 (2): 142–70.
– 2019a. 'Save the Meat for Cats: Why It's Wrong to Eat Roadkill'. *Journal of Agricultural and Environmental Ethics* 32 (2): 165–82.
– 2019b. 'Veganism, (Almost) Harm-Free Animal Flesh, and Nonmaleficence: Navigating Dietary Ethics in an Unjust World'. In *The Routledge Handbook of Animal Ethics*, edited by Bob Fischer, 555–68. Abingdon: Routledge.
– 2020. 'How to Help When It Hurts: ACT Individually (and in Groups)'. *Animal Studies Journal* 9 (1): 170–200.
Abrell, Elan. 2021. *Saving Animals*. Minneapolis: University of Minnesota Press.
Acampora, Ralph. 2004. '*Oikos* and *Domus*: On Constructive Co-Habitation with Other Creatures'. *Philosophy and Geography* 7 (2): 219–35.
Adams, Carol, and Josephine Donovan, eds. 1995. *Animals and Women*. Durham, NC: Duke University Press.
– 1996. *Beyond Animal Rights*. New York: Continuum.
– 2007. *The Feminist Care Tradition in Animal Ethics*. New York: Columbia University Press.
Ahlhaus, Svenja, and Peter Niesen. 2015. 'What Is Animal Politics? Outline of a New Research Agenda'. *Historical Social Research* 40 (4): 7–31.
Allhoff, Fritz, and Dave Monroe. 2007. 'Setting the Table: An Introduction to *Food and Philosophy*'. In *Food and Philosophy*, edited by Fritz Allhoff and Dave Monroe, 1–10. Malden: Blackwell.

Anderson, Elizabeth. 2005. 'Animal Rights and the Values of Nonhuman Life'. In *Animal Rights*, edited by Martha Nussbaum and Cass Sunstein, 277–98. Oxford: Oxford University Press.

Archer, Michael. 2011a. 'Ordering the Vegetarian Meal? There's More Animal Blood on Your Hands'. *The Conversation*. https://theconversation.com/ordering-the-vegetarian-meal-theres-more-animal-blood-on-your-hands-4659.

— 2011b. 'Slaughter of the Singing Sentients: Measuring the Morality of Eating Red Meat'. *Australian Zoologist* 35 (4): 979–82.

Barrett, Jess. 2017. 'Five Species We're Helping with Farmers and Crofters'. Royal Society for the Protection of Birds. https://community.rspb.org.uk/ourwork/b/scotland/posts/five-species-we-39-re-helping-with-farmers-and-crofters.

Beacham, Andrew M., Laura H. Vickers, and James M. Monaghan. 2019. 'Vertical Farming: A Summary of Approaches to Growing Skywards'. *The Journal of Horticultural Science and Biotechnology* 94 (3): 277–283.

Belshaw, Christopher. 2016. 'Death, Pain, and Animal Life'. In *The Ethics of Killing Animals*, edited by Rob Garner and Tatjana Višak, 32–50. Oxford: Oxford University Press.

Benke, Kurt, and Bruce Tomkins. 2017. 'Future Food-Production Systems: Vertical Farming and Controlled-Environment Agriculture'. *Sustainability: Science, Practice and Policy* 13 (1): 13–26.

Bentham, Jeremy. 1879. *An Introduction to the Principles of Morals and Legislation*. Oxford: Clarendon Press.

Blattner, Charlotte, Kendra Coulter, and Will Kymlicka, eds. 2020. *Animal Labour*. Oxford: Oxford University Press.

Blattner, Charlotte E., Sue Donaldson, and Ryan Wilcox. 2020. 'Animal Agency in Community: A Political Multispecies Ethnography of VINE Sanctuary'. *Politics and Animals* 6: 1–22.

Bobier, Christopher. 2020. 'Should Moral Vegetarians Avoid Eating Vegetables?' *Food Ethics* 5 (1).

Boisvert, Raymond D. 2014. *I Eat, Therefore I Think*. Madison, WI: Fairleigh Dickinson University Press.

Boisvert, Raymond D., and Lisa Heldke. 2016. *Philosophers at Table: On Food and Being Human*. London: Reaktion Books.

Bovenkerk, Bernice, and Jozef Keulartz, eds. 2016. *Animal Ethics in the Age of Humans*. Dordrecht: Springer.

Bovenkerk, Bernice, Eva Meijer, and Hanneke Nijland. 2020. 'Veganisme of Menselijk Diervoer? Een Niet-Antropocentrische Benadering Van Het Wereldvoedselprobleem'. In *Tien Miljard Monden*, edited by Ingrid De Zwarte and Jeroen Candel, 346–52. Amsterdam: Prometheus.

Bradshaw, Karen. 2020. *Wildlife as Property Owners*. Chicago: University of Chicago Press.

Brown, Katy. 2010. 'A Nation of Animal Lovers?' *Ethical Consumer* 123: 12–16, 34–75.

Bruckner, Donald. 2015. 'Strict Vegetarianism Is Immoral'. In *The Moral Complexities of Eating Meat*, edited by Ben Bramble and Bob Fischer, 30–47. Oxford: Oxford University Press.

Cahoone, Lawrence. 2009. 'Hunting as a Moral Good'. *Environmental Values* 18 (1): 67–89.

Callicott, J. Baird. 1992. 'Animal Liberation and Environmental Ethics: Back Together Again'. In *The Animal Liberation/Environmental Ethics Debate*, edited by Eugene Hargrove, 249–62. Albany: SUNY Press.

Cataldi, Suzanne Laba. 2002. 'Animals and the Concept of Dignity: Critical Reflections on a Circus Performance'. *Ethics and the Environment* 7 (2): 104–26.

Ceva, Emanuela, and Matteo Bonotti 2015. 'Introduction: The Political Philosophy of Food Policies, Part I: Justice, Legitimacy and Rights'. *Journal of Social Philosophy* 46 (4): 398–401.

Ciocchetti, Christopher. 2012. 'Veganism and Living Well'. *Journal of Agricultural and Environmental Ethics* 25 (3): 405–17.

Clark, Stephen R. L. 2008. '"I Knew Him by His Voice": Can Animals Be Our Friends?' *Philosophy Now* 67: 13–16.

Cochrane, Alasdair. 2009. 'Do Animals Have an Interest in Liberty?' *Political Studies* 57 (3): 660–79.

– 2010a. *An Introduction to Animals and Political Theory*. Basingstoke: Palgrave Macmillan.

– 2010b. 'Undignified Bioethics'. *Bioethics* 24 (5): 234–41.

– 2012. *Animal Rights without Liberation*. New York: Columbia University Press.

– 2014. 'Born in Chains? The Ethics of Animal Domestication'. In *The Ethics of Captivity*, edited by Lori Gruen, 156–73. Oxford: Oxford University Press.

– 2019. *Sentientist Politics*. Oxford: Oxford University Press.

– 2020. *Should Animals Have Political Rights?* Cambridge: Polity Press.

Cochrane, Alasdair, Robert Garner, and Siobhan O'Sullivan. 2018. 'Animal Ethics and the Political'. *Critical Review of International Social and Political Philosophy* 21 (2): 261–77.

Cole, Matthew, and Karen Morgan. 2013. 'Engineering Freedom? A Critique of Biotechnological Routes to Animal Liberation'. *Configurations* 21 (2): 201–29.

Cooke, Steve. 2011. 'Duties to Companion Animals'. *Res Publica* 17 (3): 261–74.
– 2017. 'Animal Kingdoms: On Habitat Rights for Wild Animals'. *Environmental Values* 26 (1): 53–72.
Coulter, Kendra. 2020. 'Toward Humane Jobs and Work-Lives for Animals'. In *Animal Labour*, edited by Charlotte E. Blattner, Kendra Coulter, and Will Kymlicka, 29–47. Oxford: Oxford University Press.
Craig, Winston J., and Ann Reed Mangels. 2009. 'Position of the American Dietetic Association: Vegetarian Diets'. *Journal of the American Dietetic Association* 109 (7): 1266–82.
Crary, Alice. 2016. *Inside Ethics*. Cambridge, MA: Harvard University Press.
– 2018. 'Ethics'. In *Critical Terms for Animal Studies*, edited by Lori Gruen, 154–68. Chicago: University of Chicago Press.
Cudworth, Erika. 2016. 'On Ambivalence and Resistance: Carnism and Diet in Multi-Species Households'. In *Meat Culture*, edited by Annie Potts, 222–42. Leiden: Brill.
Davis, Chris. 2018. 'Pig Farm Begs Question: How High the Sty? Call the Elevator, Please'. *China Daily*. http://www.chinadaily.com.cn/a/201805/31/WS5b0fa5baa31001b82571d762.html.
Davis, Steven L. 2003. 'The Least Harm Principle May Require that Humans Consume a Diet Containing Large Herbivores, Not a Vegan Diet'. *Journal of Agricultural and Environmental Ethics* 16 (4): 387–94.
Deckers, Jan. 2016. *Animal (De)Liberation*. London: Ubiquity Press.
Derham, Tristan, and Freya Mathews. 2020. 'Elephants as Refugees'. *People and Nature* 2 (1): 103–10.
Despommier, Dickson. 2009. 'The Rise of Vertical Farms'. *Scientific American* 301 (5): 80–7.
– 2013. 'Farming Up the City: The Rise of Urban Vertical Farms'. *Trends in Biotechnology* 31 (7): 388–9.
Diamond, Cora. 1978. 'Eating Meat and Eating People'. *Philosophy* 53 (206): 465–79.
Dieterle, J.M., ed. 2015. *Just Food*. Lanham: Rowman and Littlefield.
Dodd, Sarah A.S., Cate Dewey, Deep Khosa, and Adronie Verburgghe. 2021. 'A Cross-Sectional Study of Owner-Reported Health in Canadian and American Cats Fed Meat- and Plant-Based Diets'. *BMC Veterinary Research* 17 (53): 1–16.
Dolgert, Stefan. 2018. 'Vegetarian Republic: Pythagorean Themes in Plato's *Republic*'. *Proceedings of the XXIII World Congress of Philosophy* 2 (2): 83–8.
Dombrowski, Daniel A. 1984. 'Was Plato a Vegetarian?' *Apeiron* 18 (1): 1–9.
Donaldson, Sue, and Will Kymlicka. 2011. *Zoopolis*. Oxford: Oxford University Press.

- 2013a. 'A Defense of Animal Citizens and Sovereigns'. *Law, Ethics and Philosophy* 1: 143–60.
- 2013b. 'A Reply to Svärd, Nurse, and Ryland'. *Journal of Animal Ethics* 3 (2): 208–19.
- 2014. 'Animal Rights, Multiculturalism, and the Left'. *Journal of Social Philosophy* 45 (1): 116–35.
- 2015a. 'Farmed Animal Sanctuaries: The Heart of the Movement?' *Politics and Animals* 1: 50–74.
- 2015b. 'Interspecies Politics: Reply to Hinchcliffe and Ladwig'. *Journal of Political Philosophy* 23 (3): 321–44.
- 2016. 'Between Wildness and Domestication: Rethinking Categories and Boundaries in Response to Animal Agency'. In *Animal Ethics in the Age of Humans*, edited by Bernice Bovenkerk and Jozef Keulartz, 225–39. Dordrecht: Springer.

Dussault, Antoine C., and Élise Desaulniers. 2019. 'Natural Food'. In *Encyclopedia of Food and Agricultural Ethics: Second Edition*, edited by Paul B. Thompson and David M. Kaplan, 1871–80. Dordrecht: Springer.

Dutkiewicz, Jan, and Elan Abrell. 2021. 'Sanctuary to Table Dining: Cellular Agriculture and the Ethics of Cell Donor Animals'. *Politics and Animals* 7: 1–15.

Dworkin, Ronald. 1984. 'Rights as Trumps'. In *Theories of Rights*, edited by Jeremy Waldron, 153–67. Oxford: Oxford University Press.

Evans, Matthew. 2019. *On Eating Meat*. Sydney: Murdoch Books.

Everett, Jennifer. 2001. 'Environmental Ethics, Animal Welfarism, and the Problem of Predation: A Bambi Lover's Respect for Nature'. *Ethics and the Environment* 6 (1): 42–67.

Feinberg, Joel. 1980. *Rights, Justice and the Bounds of Liberty*. Princeton, NJ: Princeton University Press.

Ferguson, Benjamin, and Christopher Thompson. 2020. 'Why Buy Local?' *Journal of Applied Philosophy* 38 (1): 104–20.

Fischer, Bob. 2016a. 'Bugging the Strict Vegan'. *Journal of Agricultural and Environmental Ethics* 29 (2): 255–63.
- 2016b. 'What If Klein and Barron Are Right about Insect Sentience?' *Animal Sentience* 115: 1–7.
- 2018. 'Arguments for Consuming Animal Products'. In *The Oxford Handbook of Food Ethics*, edited by Anne Barnehill, Tyler Doggett, and Mark Budofson, 241–66. Oxford: Oxford University Press.
- 2019a. 'Meat: Ethical Considerations'. In *Encyclopedia of Food and Agricultural Ethics: Second Edition*, edited by David M. Kaplan and Paul B. Thompson, 1788–94. Dordrecht: Springer.
- 2019b. *The Ethics of Eating Animals*. Abingdon: Routledge.

Fischer, Bob, and Andy Lamey. 2018. 'Field Deaths in Plant Agriculture.' *Journal of Agricultural and Environmental Ethics* 31 (4): 409–28.
Fischer, Bob, and Josh Milburn. 2019. 'In Defence of Backyard Chickens.' *Journal of Applied Philosophy* 36 (1): 108–23.
Fischer, Bob, and Burkay Ozturk. 2017. 'Facsimiles of Flesh.' *Journal of Applied Philosophy* 34 (4): 489–97.
Francione, Gary L. 2007. *Introduction to Animal Rights*. Philadelphia: Temple University Press.
– 2014. 'Profile of a Modern "Animal Activist": Jenna Woginrich.' Abolitionist Approach. http://www.abolitionistapproach.com/profile-modern-animal-activist-jenna-woginrichl-activist/.
Francione, Gary L., and Anna Charlton. 2013. *Eat Like You Care*. Logan: Exempla Press.
– 2015. *Animal Rights: The Abolitionist Approach*. Logan: Exempla Press.
– 2016. 'Veganism without Animal Rights.' In *The Routledge Handbook of Food Ethics*, edited by Mary C. Rawlinson and Caleb Ward, 294–304. Abingdon: Routledge.
Francione, Gary L., and Robert Garner. 2010. *The Animal Rights Debate*. New York: Columbia University Press.
Frey, Raymond G. 1977. 'Animal Rights.' *Analysis* 37 (4): 186–9.
Fröding, Barbro, and Martin Peterson. 2011. 'Animal Ethics Based on Friendship.' *Journal of Animal Ethics* 1 (1): 58–69.
Garner, Robert. 2013. *A Theory of Justice for Animals*. Oxford: Oxford University Press.
Garner, Robert, and Yewande Okuleye. 2020. *The Oxford Group and the Emergence of Animal Rights*. Oxford: Oxford University Press.
Gibert, Martin, and Elise Desaulniers. 2019. 'Carnism.' In *Encyclopedia of Food and Agricultural Ethics: Second Edition*, edited by David Kaplan and Paul B. Thompson, 372–8. Dordrecht: Springer.
Gillen, Jed. 2003. *Obligate Carnivore*. Seattle: Steinhoist Books.
Giroux, Valéry. 2016. 'Animals Do Have an Interest in Liberty.' *Journal of Animal Ethics* 6 (1): 20–43.
Godlovitch, Stanley, Rosalind Godlovitch, and John Harris, eds. 1971. *Animals, Men and Morals*. London: Victor Gollancz.
Goodin, Robert E., Carole Pateman, and Roy Pateman. 1997. 'Simian Sovereignty.' *Political Theory* 25 (6): 821–49.
Goodnick, Liz. 2015. 'Limits on Locavorism.' In *Just Food*, edited by J.M. Dieterle, 195–212. London: Rowman and Littlefield International.
Gottlieb, Robert, and Anupama Joshi. 2010. *Food Justice*. Cambridge, MA: The MIT Press.

Gray, Christina M., Rance K. Sellon, and Lisa M. Freeman. 2004. 'Nutritional Adequacy of Two Vegan Diets for Cats'. *Journal of the American Veterinary Medical Association* 225 (11): 1670–5.

Gruen, Lori. 2011. *Ethics and Animals*. Cambridge: Cambridge University Press.

– 2014. 'Dignity, Captivity, and an Ethics of Sight'. In *The Ethics of Captivity*, edited by Lori Gruen, 231–47. Oxford: Oxford University Press.

Gruen, Lori, and Robert C. Jones. 2015. 'Veganism as an Aspiration'. In *The Ethical Complexities of Eating Meat*, edited by Bob Fischer and Ben Bradley, 153–71. Oxford: Oxford University Press.

Hadley, John. 2006. 'The Duty to Aid Nonhuman Animals in Dire Need'. *Journal of Applied Philosophy* 23 (4): 445–51.

– 2016. *Animal Property Rights: A Theory of Habitat Rights for Wild Animals*. Lanham: Lexington.

Harper, A. Breeze, ed. 2010. *Sistah Vegan*. New York: Lantern Books.

Healey, Richard, and Angie Pepper. 2021. 'Interspecies Justice: Agency, Self-Determination, and Assent'. *Philosophical Studies* 178: 1223–43.

Hobbes, Thomas. 1996. *Leviathan*, edited by Richard Tuck. Cambridge: Cambridge University Press.

Hooley, Dan, and Nathan Nobis. 2016. 'A Moral Argument for Veganism'. In *Philosophy Comes to Dinner*, edited by Andrew Chignell, Terence Cuneo, and Matthew C. Halteman, 92–108. Abingdon: Routledge.

Hopkins, Patrick D., and Austin Dacey. 2008. 'Vegetarian Meat: Could Technology Save Animals and Satisfy Meat Eaters?' *Journal of Agricultural and Environmental Ethics* 21 (6): 579–96.

Horta, Oscar. 2010. 'Debunking the Idyllic View of Natural Processes: Population Dynamics and Suffering in the Wild'. *Telos* 17 (1): 73–90.

– 2011. 'The Ethics of the Ecology of Fear against the Nonspeciesist Paradigm: A Shift in the Aims of Intervention in Nature'. *Between the Species* 13 (10): 163–87.

– 2013. '*Zoopolis*, Intervention, and the State of Nature'. *Law, Ethics and Philosophy* 1 (1): 113–25.

– 2017. 'Animal Suffering in Nature: The Case for Intervention'. *Environmental Ethics* 39 (3): 261–79.

Irvine, Leslie. 2013. *My Dog Always Eats First*. Boulder, CO: Lynne Rienner Publishers.

Jamieson, Dale. 2008. 'The Rights of Animals and the Demands of Nature'. *Environmental Values* 17 (2): 181–99.

Jenkins, Simon. 2016. 'Food, Welfare, and Agriculture: A Complex Picture'. In *The Routledge Handbook of Food Ethics*, edited by Mary C. Rawlinson and Caleb Ward, 274–83. Abingdon: Routledge.

Johannsen, Kyle. 2017. 'Animal Rights and the Problem of r-Strategists'. *Ethical Theory and Moral Practice* 20 (2): 333–45.
– 2021. *Wild Animal Ethics*. Abingdon: Routledge.
Jordan, Jeff. 2001. 'Why Friends Shouldn't Let Friends Be Eaten: An Argument for Vegetarianism'. *Social Theory and Practice* 27 (2): 309–22.
Joy, Melanie. 2010. *Why We Love Dogs, Eat Pigs, and Wear Cows*. San Francisco: Conari Press.
Kaiser, Matthias, and Anne Algers. 2016. 'Food Ethics: A Wide Field in Need of Dialogue'. *Food Ethics* 1 (1): 1–7.
Kalantari, Fatemeh, Osman Mohd Tahir, Raheleh Akbari Joni, and Ezaz Fatemi. 2018. 'Opportunities and Challenges in Sustainability of Vertical Farming: A Review'. *Journal of Landscape Ecology* 11 (1): 35–60.
Kant, Immanuel. 1997. 'Of Duties to Animals and Spirits'. In *Lectures on Ethics*, translated by Peter Heath, edited by Peter Heath and J.B. Schneewind, 212–13. Cambridge: Cambridge University Press.
– 2003. *To Perpetual Peace*, translated by Ted Humphrey. Indianapolis: Hackett.
Kaplan, David. 2012. 'Introduction: The Philosophy of Food'. In *The Philosophy of Food*, edited by David Kaplan, 1–23. Berkeley: University of California Press.
Kazez, Jean. 2018. 'The Taste Question in Animal Ethics'. *Journal of Applied Philosophy* 35 (4): 661–74.
Keulartz, Jozef. 2016. 'Should the Lion Eat Straw Like the Ox? Animal Ethics and the Predation Problem'. *Journal of Agricultural and Environmental Ethics* 29 (5): 813–34.
Kianpour, Connor Kayhan. 2020. 'Cetacean Property: A Hegelish Account of Nonhuman Property'. *Politics and Animals* 6: 23–36.
Knight, Andrew, and Madelaine Leitsberger. 2016. 'Vegetarian versus Meat-Based Diets for Companion Animals'. *Animals* 6 (9).
Korsmeyer, Carolyn. 2002. *Making Sense of Taste*. Ithaca, NY: Cornell University Press.
Kotzen, Benz, Maurício Gustavo Coelho Emerenciano, Navid Moheimani, and Gavin M. Burnell. 2019. 'Aquaponics: Alternative Types and Approaches'. In *Aquaponics Food Production Systems*, edited by Simon Goddek, Alyssa Joyce, Benz Kotzen, and Gavin M. Burnell, 301–30. Cham: SpringerOpen.
Ladwig, Bernd. 2015. 'Against Wild Animal Sovereignty: An Interest-Based Critique of *Zoopolis*'. *Journal of Political Philosophy* 23 (3): 282–301.
Lamey, Andy. 2007. 'Food Fight! Davis versus Regan on the Ethics of Eating Beef'. *Journal of Social Philosophy* 38 (2): 331–48.
– 2019. *Duty and the Beast*. Cambridge: Cambridge University Press.

Lomasky, Loren. 2013. 'Is It Wrong to Eat Animals?' *Social Philosophy and Policy* 30 (1–2): 177–200.

Lorenzini, Daniele. 2020. 'The Definition of Nonhuman Animal Euthanasia.' *Animal Studies Journal* 9 (2): 1–20.

MacDonald, M.L., Q.R. Rogers, and J.G. Morris. 1984. 'Nutrition of the Domestic Cat, a Mammalian Carnivore.' *Annual Review of Nutrition* 4: 521–62.

Marra, Peter P., and Chris Santella. 2016. *Cat Wars*. Princeton, NJ: Princeton University Press.

Matheny, Gaverick. 2003. 'Least Harm: A Defense of Vegetarianism from Steven Davis's Omnivorous Proposal.' *Journal of Agricultural and Environmental Ethics* 16 (5): 505–11.

McMahan, Jeff. 2010. 'The Meat Eaters.' *New York Times*. http://opinionator.blogs.nytimes.com/2010/09/19/the-meat-eaters/.

– 2015. 'The Moral Problem of Predation.' In *Philosophy Comes to Dinner*, edited by Andrew Chignell, Terence Cuneo, and Matthew C. Halteman, 268–93. Abingdon: Routledge.

McWilliams, James E. 2009. *Just Food*. New York: Little, Brown and Company.

Meijer, Eva. 2019. *When Animals Speak*. New York: New York University Press.

Meyer, Michael. 2001. 'The Simple Dignity of Sentient Life: Speciesism and Human Dignity.' *Journal of Social Philosophy* 32 (2): 115–26.

Meyers, Chris D. 2013. 'Why It Is Morally Good to Eat (Certain Kinds of) Meat: The Case for Entomophagy.' *Southwest Philosophy Review* 29 (1): 119–26.

Michel, Kathryn. 2006. 'Unconventional Diets for Dogs and Cats.' *Veterinary Clinics of North America: Small Animal Practice* 36 (6): 1269–81.

Midgley, Mary. 1983. *Animals and Why They Matter*. Athens: University of Georgia Press.

Milburn, Josh. 2015a. 'Not Only Humans Eat Meat: Companions, Sentience, and Vegan Politics.' *Journal of Social Philosophy* 46 (4): 449–62.

– 2015b. 'Rabbits, Stoats and the Predator Problem: Why a Strong Animal Rights Position Need Not Call for Human Intervention to Protect Prey from Predators.' *Res Publica* 21 (3): 273–89.

– 2016a. 'Animal Rights and Food: Beyond Regan, Beyond Vegan.' In *The Routledge Handbook of Food Ethics*, edited by Mary C. Rawlinson and Caleb Ward, 284–93. Abingdon: Routledge.

– 2016b. 'Chewing Over *In Vitro* Meat: Animal Ethics, Cannibalism and Social Progress.' *Res Publica* 22 (3): 249–65.

– 2016c. 'Nonhuman Animals and Sovereignty: On *Zoopolis*, Failed States and Institutional Relationships with Free-Living Animals'. In *Intervention or Protest*, edited by Andrew Woodhall and Gabriel Garmendia da Trindade, 183–212. Wilmington, DE: Vernon Press.
– 2017a. 'Nonhuman Animals as Property Holders: An Exploration of the Lockean Labour-Mixing Account'. *Environmental Values* 26 (5): 629–48.
– 2017b. 'The Animal Lovers' Paradox? On the Ethics of "Pet Food"'. In *Pets and People*, edited by Christine Overall, 187–202. Oxford: Oxford University Press.
– 2018. 'Death-Free Dairy? The Ethics of Clean Milk'. *Journal of Agricultural and Environmental Ethics* 31 (2): 261–79.
– 2019a. 'Pet Food: Ethical Issues'. In *Encyclopedia of Food and Agricultural Ethics: Second Edition*, edited by Paul B. Thompson and David M. Kaplan, 1967–73. Dordrecht: Springer.
– 2019b. 'Sentientist Politics Gone Wild'. *Politics and Animals* 5: 19–24.
– 2019c. 'Vegetarian Eating'. In *Handbook of Eating and Drinking*, edited by Herbert L. Meiselman. Dordrecht: Springer.
– 2020a. 'A Novel Case for Vegetarianism? Zoopolitics and Respect for Animal Corpses'. *Animal Studies Journal* 9 (2): 240–59.
– 2020b. 'Should Vegans Compromise?' *Critical Review of International Social and Political Philosophy*. doi:10.1080/13698230.2020.1737477.
Milburn, Josh, and Bob Fischer. 2021. 'The Freegan Challenge to Veganism'. *Journal of Agricultural and Environmental Ethics* 34 (17): 1–19.
Milburn, Josh, and Sara Van Goozen. 2021. 'Counting Animals in War: First Steps towards an Inclusive Just-War Theory'. *Social Theory and Practice* 47 (4): 657–85.
Mill, John Stuart. 1867. 'Dr. Whewell on Moral Philosophy'. In *Discussions Political, Philosophical and Historical*, vol. 2, 450–509. London: Longmans, Greene, Reader, and Dyer.
Miller, John. 2012. 'In Vitro Meat: Power, Authenticity and Vegetarianism'. *Journal for Critical Animal Studies* 10 (4): 41–63.
Milligan, Tony. 2010. *Beyond Animal Rights*. London: Continuum.
– 2015. 'The Political Turn in Animal Rights'. *Politics and Animals* 1: 6–15.
Nagel, Thomas. 1974. 'What Is It Like to Be a Bat?' *Philosophical Review* 83 (4): 435–50.
Narveson, Jan. 1999. *Moral Matters*. Peterborough: Broadview Press.
Nestle, Marion. 2008. *Pet Food Politics*. Berkeley: University of California Press.
Nestle, Marion, and Malden C. Nesheim. 2010. *Feed Your Pet Right*. New York: Free Press.

Nozick, Robert. 1974. *Anarchy, State, and Utopia*. New York: Basic Books.
– 1981. *Philosophical Explanations*. Oxford: Oxford University Press.
Nussbaum, Martha. 2006. *Frontiers of Justice*. Cambridge, MA: Harvard University Press.
O'Connor, Terry. 2013. *Animals as Neighbors*. East Lansing: Michigan State University Press.
Painter, Corinne M. 2013. 'The Vegetarian Polis: Just Diet in Plato's Republic and in Ours.' *Journal of Animal Ethics* 3 (2): 121–32.
Palmer, Clare. 2003. 'Placing Animals in Urban Environmental Ethics.' *Journal of Social Philosophy* 34 (1): 64–78.
– 2010. *Animal Ethics in Context*. New York: Columbia University Press.
– 2020. 'The *Laissez-Faire* View: Why We're Not Normally Required to Assist Wild Animals.' In *The Routledge Handbook of Animal Ethics*, edited by Bob Fischer, 444–54. Abingdon: Routledge.
– 2021. 'Should We Provide the Bear Necessities? Climate Change, Polar Bears and the Ethics of Supplemental Feeding.' In *Animals in Our Midst*, edited by Bernice Bovenkirk and Josef Keulartz, 377–98. Leiden: Springer.
Pierce, Jessica. 2016. *Run, Spot, Run*. Chicago: University of Chicago Press.
Platt, Peter. 2007. 'Vertical Farming: An Interview with Dickson Despommier.' *Gastronomica* 7 (3): 80–7.
Pollan, Michael. 2006. *The Omnivore's Dilemma*. New York: Penguin.
Porphyry. 2000. *On Abstinence from Killing Animals*, translated by Gillian Clark. London: Duckworth.
Posner, Richard A. 2005. 'Animal Rights: Legal, Philosophical and Pragmatic Perspectives.' In *Animal Rights*, edited by Martha Nussbaum and Cass Sunstein, 51–70. Oxford: Oxford University Press.
Rachels, James. 1990. *Created from Animals*. Oxford: Oxford University Press.
Regan, Tom. 1976. 'McCloskey on Why Animals Cannot Have Rights.' *Philosophical Quarterly* 26 (104): 251–7.
– 1984. *The Case for Animal Rights*. London: Routledge and Kegan Paul.
– 2004. *The Case for Animal Rights* (updated ed.). Berkeley: University of California Press.
Robaina, Lidia, Juhani Pirhonen, Elena Mente, Javier Sánchez, and Neill Goosen. 2019. 'Fish Diets in Aquaponics.' In *Aquaponics Food Production Systems*, edited by Simon Goddek, Alyssa Joyce, Benz Kotzen, and Gavin M. Burnell, 333–52. Cham: SpringerOpen.
Rothgerber, Hank. 2013. 'A Meaty Matter: Pet Diet and the Vegetarian's Dilemma.' *Appetite* 68 (1): 76–82.
– 2014. 'Carnivorous Cats, Vegetarian Dogs, and the Resolution of the Vegetarian's Dilemma.' *Anthrozoös* 27 (4): 485–98.

Rousseau, Jean-Jacques. 2002. *Discourse on the Origin and Foundations of Inequality among Mankind*. In *The Social Contract and the First and Second Discourses*, edited by Susan Dunn, 69–148. New Haven, CT: Yale University Press.

Rowlands, Mark. 2009. *Animal Rights*. Basingstoke: Palgrave Macmillan.

Sagoff, Mark. 1984. 'Animal Liberation and Environmental Ethics: Bad Marriage, Quick Divorce'. *Osgoode Hall Law Journal* 22: 297–307.

Salt, Henry S. 1980. *Animals' Rights*. Clarks Summit: Society for Animal Rights.

Sandøe, Peter, Sandra Corr, and Clare Palmer. 2015. *Companion Animal Ethics*. Hoboken, NJ: Wiley-Blackwell.

Sapontzis, Steve. 1987. *Morals, Reason and Animals*. Philadelphia: Temple University Press.

Schaefer, G. Owen, and Julian Savulescu. 2014. 'The Ethics of Producing In Vitro Meat'. *Journal of Applied Philosophy* 31 (2): 188–202.

Schedler, George. 2005. 'Does Ethical Meat Eating Maximise Utility?' *Social Theory and Practice* 31 (4): 499–511.

Schlottmann, Christopher, and Jeff Sebo. 2018. *Food, Animals, and the Environment*. Abingdon: Routledge.

Scotton, Guy. 2017. 'Duties to Socialise with Domesticated Animals: Farmed Animal Sanctuaries as Frontiers of Friendship'. *Animal Studies Journal* 6 (2): 86–108.

Sebo, Jeff. 2021. *Animal Ethics in a Human World*. Oxford: Oxford University Press.

Siipi, Helena. 2008. 'Dimensions of Naturalness'. *Ethics and the Environment* 13 (1): 71–103.

– 2013. 'Is Natural Food Healthy?' *Journal of Agricultural and Environmental Ethics* 26 (4): 797–812.

Silvermintz, Daniel. 2019. 'Plato and Food'. In *Encyclopedia of Food and Agricultural Ethics: Second Edition*, edited by David Kaplan and Paul B. Thompson, 2001–7. Dordrecht: Springer.

Singer, Peter. 1974. 'All Animals Are Equal'. *Philosophic Exchange* 5 (1): 103–16.

– 1975. *Animal Liberation*. New York: HarperCollins.

– 1995. *Animal Liberation* (2nd ed.). London: Pimlico.

– 1999. 'A Response'. In *Singer and His Critics*, edited by Dale Jamieson, 269–335. Oxford: Blackwell.

Smith, Kimberly. 2012. *Governing Animals*. Oxford: Oxford University Press.

Stănescu, Vasile. 2016. 'Beyond Happy Meat: The (Im)Possibilities of "Humane", "Local" and "Compassionate" Meat'. In *The Future of Meat without Animals*, edited by Brianne Donaldson and Christopher Carter, 133–54. London: Rowman and Littlefield International.

Szymanski, Ileana. 2016. 'What Is Food? Networks, Not Commodities'. In *The Routledge Handbook of Food Ethics*, edited by Mary C. Rawlinson and Caleb Ward, 7–15. Abingdon: Routledge.

Taylor, Angus. 2019. 'Should Vegans Eat Meat to Kill Fewer Animals?' Animals and Society Research Initiative. https://onlineacademiccommunity.uvic.ca/asri/2020/05/28/should-vegans-eat-meat-to-kill-fewer-animals/.

Telfer, Elizabeth. 1995. 'Hospitableness'. *Philosophical Papers* 24 (3): 183–96.

– 1996. *Food for Thought*. Abingdon: Routledge.

Thompson, Paul B. 2016. 'The Emergence of Food Ethics'. *Food Ethics* 1: 61–74.

Tomasik, Brian. 2015. 'The Importance of Wild-Animal Suffering'. *Relations. Beyond Anthropocentrism* 3 (2): 133–52.

– 2019. 'How Many Wild Animals Are There?' Essays On Reducing Suffering. https://reducing-suffering.org/how-many-wild-animals-are-there/.

Townley, Cynthia. 2011. 'Animals as Friends'. *Between the Species* 13 (10): 45–59.

– 2017. 'Friendship with Companion Animals'. In *Pets and People*, edited by Christine Overall, 21–34. Oxford: Oxford University Press.

Tuomisto, Hanna L. 2019. 'Vertical Farming and Cultured Meat: Immature Technologies for Urgent Problems'. *One Earth* 1 (3): 275–6.

Turner, Susan. 2005. 'Beyond Viande: The Ethics of Faux Flesh, Fake Fur and Thriftshop Leather'. *Between the Species* 13 (5): 1–13.

Valentini, Laura. 2014. 'Canine Justice: An Associative Account'. *Political Studies* 62 (1): 37–52.

Varner, Gary. 1998. *In Nature's Interests?* Oxford: Oxford University Press.

Wadiwel, Dinesh. 2015. *The War against Animals*. Leiden: Brill.

Wakefield, Lorelei A., Frances S. Shofer, and Kathryn E. Michel. 2006. 'Evaluation of Cats Fed Vegetarian Diets and the Attitudes of their Caregivers'. *Journal of the American Veterinary Medical Association* 229 (1): 70–3.

Ward, Ernie, Alice Oven, and Ryan Bethencourt. 2020. *The Clean Pet Food Revolution*. New York: Lantern Books.

Wayne, Katherine. 2013a. 'Just Flourishing: The Plausibility of Selective Extinctionism'. Unpublished manuscript.

– 2013b. 'Permissible Use and Interdependence: Against Principled Veganism'. *Journal of Applied Philosophy* 30 (2): 160–75.

Weele, Cor Van Der, and Clemens Driessen. 2013. 'Emerging Profiles for Cultured Meat; Ethics through and as Design'. *Animals* 3 (3): 647–62.

Woginrich, Jenna. 2014. 'An Open Letter to Angry Vegetarians'. Cold Antler Farm. http://coldantlerfarm.blogspot.co.uk/2014/07/an-open-letter-to-angry-vegetarians.html.

Wrenn, Corey Lee. 2016. *A Rational Approach to Animal Rights*. Basingstoke: Palgrave MacMillan.

– 2018. 'How to Help When It Hurts? Think Systematic.' *Animal Studies Journal* 7 (1): 149–79.

Wrye, Jen. 2015. '"Deep Inside Dogs Know What They Want": Animality, Affect, and Killability in Commercial Pet Foods.' In *Economies of Death*, edited by Tish Lopez and Kathryn Gillespie, 95–114. Abingdon: Routledge.

Zamir, Tzachi. 2007. *Ethics and the Beast*. Princeton, NJ: Princeton University Press.

Zwart, Hub. 2000. 'A Short History of Food Ethics.' *Journal of Agricultural and Environmental Ethics* 12 (2): 113–36.

Index

page numbers with (t) refer
to tables.

Abbate, Cheryl, 37, 137–8, 142–7, 154, 198n12
abolitionism, 7–8, 39–40, 43–4
aeroponics, 123, 129. *See also* vertical agriculture
affect. *See* relational framework
agency. *See* choice and autonomy; moral agency
agriculture: history of, 191n7; overproduction, 119; rewilded habitats, 125, 132–4, 161, 196n18; as 'unnatural', 59–60; vertical farms, 122–3. *See also* animal agriculture; arable agriculture; cellular agriculture; cultivated meat; food production system; vertical agriculture
Algers, Anne, 14, 185n12
Allhoff, Fritz, 185n12
Anderson, Elizabeth, 190n5
animal agriculture: about, 23–7, 179, 186n18; burger vegans' critiques, 107–18; collective responsibilities, 44–5, 128–30, 143–4; cultural meanings, 25, 27, 30, 61–2; doctrine of double effect, 117–18; food for, 112, 194n2; food for WRCS, 143–4, 146–7; food justice, 54–5, 83–4, 186n14; intentions, 117–18; interest-based rights, 26–7, 128; milk production, 24, 44, 146, 197n6; problem of carnivory, 22–3, 28–32, 47–9; sanctuaries for farmed animals, 139–40, 153–5; statistics, 23; suffering and death, 23–7, 30–2, 116–17; veganism as moral necessity, 19–20, 23–7; vertical agriculture, 128–30. *See also* carnivores; fish and 'sea food'; suffering and death
animal companions. *See* companion animals
animal ethics: about, 5–8, 16–17, 20; care ethics, 7–8, 32, 52; consequentialism, 8, 10, 117, 161–2; feminist critiques, 7–8, 11; history of, 6–8, 183–5nn2–10; interest-based rights, 26–7; key questions, 3–5; philosophy of food, 16–17; precautionary principle, 45–6; *prima facie* and concrete rights, 26–7; problem of carnivory, 22–3, 28–32, 47–9; sentience, 45–7; utilitarianism, 6–10; value theory, 7–8; veganism

as moral baseline, 39–40; without rights, 45–9. *See also* abolitionism; carnivores, problem of carnivory; food and philosophy; interest-based rights; moral philosophy; normative obligations; political philosophy; *prima facie* and concrete rights; rights-based approaches; sentience; utilitarianism

animal friends. *See* animal neighbours, friends

animal-human interaction. *See* animals; humans; normative obligations; relational framework

animal neighbours: about, 18, 82–4, 181; analogy to human neighbours, 88–9, 91–3, 96–7, 105; collective responsibilities, 93–4, 98–9, 105–6, 181; concept of, 84–8; dependency, 83, 87, 92–5; harms to humans, 100, 105–6; imperfect/perfect duties, 89, 90, 96, 193n7; key questions, 3–5, 83; problem of carnivory, 83; recommendations, 181; relational framework, 83, 86–7, 98–9, 105–6, 192n1; risk of harm to, 95–100. *See also* birds; carnivores, problem of carnivory; hospitality

animal neighbours, categories: about, 82–7; based on feeding relationships, 28–9, 84–8; commensal animals, 84–5, 192n1; vs companions, 92; contact zone, 82–3; enemies/pests/vermin, 84, 87, 192n4; friends vs foes, 84–8, 90, 181; ignored/disinterested, 84, 88, 97, 103; liminal animals, 82–4, 99–100, 134; parasites, 192n1, 192n4; relational framework, 86–7; synanthropes, 84–5; vs wild animals, 85. *See also* animal neighbours, foes; animal neighbours, friends; biological classifications; carnivores; herbivores; omnivores

animal neighbours, feeding: about, 83–4, 87–92, 181; dependency, 87–8, 92–5, 181; duty to feed, 83, 87–90, 97; food sources, 83–4; hospitality, 18, 88–92, 96–7, 106. *See also* hospitality

animal neighbours, foes: about, 85–7, 90–2, 106, 181; animal thieves as, 108–9, 119–20; as category, 85–8, 192n4; denial of hospitality, 18, 87, 90–2, 97, 101–6, 119–20; denial of protection of, 97, 101–4; vs friends, 84–8, 90, 181; negative rights, 90–1, 102–4. *See also* animal neighbours, categories; animal thieves

animal neighbours, friends: about, 84–8, 181; analogies to humans, 84, 86, 88–9, 91–3, 105; collective responsibilities, 98–9; dependency, 92–5; duty to feed, 87–92; vs foes, 84–8, 90, 181; harms to humans, 100, 105–6; hospitality, 18, 89–92, 95–7, 106; predator problem, 83, 98–100; relational framework, 86–7; safety responsibilities, 95–7. *See also* animal neighbours, categories; birds

animal refugees: about, 19, 135–8; climate change effects, 19, 138, 172–5, 177; collective responsibilities, 138, 143, 154–5; displaced animals, 138; from institutions, 142–3; predator problem, 137; sanctuaries, 135–7; in WRCS, 135–6. *See also* sanctuaries; wild animals; WRCS (wildlife rehabilitation centres);

WRCS (wildlife rehabilitation centres), predator problem
animal thieves: about, 18–19, 107–9, 134; dependency, 132–4, 195n16; as foes, 108–9, 119–20; as friends, 133–4; normative obligations, 18, 131–4; as wild animals, 114, 133. *See also* animal neighbours, foes
animal thieves, harms: burger vegans' critiques, 107–11, 113–18, 134; harvesting suffering, 109–10, 120; statistics, 113–16; by vertical farms, 125–6, 130–4
animal thieves, prevention of harms: about, 18–19, 108, 117–21, 134; collective responsibilities, 118–22, 126–34; methods, 108, 118–22; normative obligations, 18–19; rewilded habitats, 125, 132–4, 196n18; vertical farms, 125–6, 131–4. *See also* burger vegans; vertical agriculture
animals: animal testing, 129, 142, 170; breeding of, 24, 39, 53, 71, 72–3, 76; classifications by feeding practices, 28–9, 34–6, 84–6; interest in continued life, 26, 33, 161, 178, 199n5; key questions, 3–5; population control, 118–19; reproduction, 159–61; sentience, 26, 45–7, 146–7; speciesism, 6, 56, 79, 164, 199n4; suffering and death, 23–7, 116–17, 159–61; terminology, 183n1. *See also* animal agriculture; animal ethics; biological classifications; carnivores; humans; interest-based rights; moral agency; relational framework; sanctuaries; sentience; suffering and death
anti-vegans, 107. *See also* burger vegans

aquaponics, 129, 195n15. *See also* vertical agriculture
Aquinas, St Thomas, 183n2
arable agriculture: about, 181; crops, 112, 194n2; food justice, 54–5, 83–4, 186n14; land-use policies, 119–20; rewilded habitats, 125, 132–4, 196n18; vertical farms, 122–5, 130. *See also* veganism and vegetarianism; vertical agriculture
arable agriculture, harms to animals: about, 107–11, 115–23, 130; burger vegans' critiques, 107–11, 113–17; collective responsibilities, 118–22, 126–30; doctrine of double effect, 117–18; during harvests, 109–10, 116; intentions, 117–18; methods to reduce, 109, 118–22; population control, 118–19, 130; statistics, 113–16; vertical farms, 122–3, 125–6. *See also* animal thieves, harms; burger vegans
Archer, Michael, 114–15, 126
Aristotle, 190n5
autonomy. *See* choice and autonomy; moral agency

BARF (natural foods) diet, 57–8
bears, 160–1, 173–5, 190n6, 200nn12–13
belonging. *See* relational framework
Belshaw, Chris, 189n2
beneficence, duty of, 19, 158, 163–71, 176–7, 199n4
Benke, Kurt, 127
Bentham, Jeremy, 184n3
biological classifications: about, 28–9, 34–6; based on feeding relationships, 28; continuum of omnivores and carnivores, 34–6; facultative vs obligate carnivores,

34–5, 141–2, 141(t), 144, 188n9; herbivores, 28–9, 34–6, 141–2, 141(t), 156; omnivores, 28–9, 34–6, 188n12. *See also* carnivores; invertebrates; omnivores

birds: collective responsibilities, 10, 93–4, 98–9, 120–1; corncrakes case, 107, 120–1; dependency, 93–4, 97; as friends, 81, 84, 86–8, 90–2, 93–4, 96–7; negative rights, 102–4; population control, 94; predator problem, 95–8, 100, 102–4, 106; relational framework, 86–7; reproduction, 159–61; safety responsibilities, 95–8, 102–4; sparrows/sparrowhawk case, 102–4; suffering and death, 159–61; as thieves, 107; Trafalgar pigeons case, 3–4, 82, 94, 106; vultures, 141–2, 141(t). *See also* animal neighbours; animal thieves; chickens; gardens; pigeons

blackbirds, 86, 100, 159–60

Boisvert, Raymond D., 186n17, 192n4, 193nn8–9

Bonotti, Matteo, 14

breeding of animals, 24, 39, 53, 71–3, 76

Bruckner, Donald, 36, 115

burger vegans: about, 107–22, 134, 181; critiques of veganism, 107–17; defined, 194n1; doctrine of double effect, 117–18; harms to animals, 107–10, 114, 116; non-vegan food for, 107–8, 110–15, 116; statistics, 111, 113–16; vertical farms as response to, 108, 126, 133–4. *See also* animal thieves, harms; animal thieves, prevention of harms; vertical agriculture

Callicott, J. Baird, 58

cannibalism, 43, 188n15, 189n19

care ethics, 7–8, 32, 52. *See also* relational framework

carnism, 28, 58, 187n6

carnivores: about, 17–18, 21–3, 28–31, 47–9; as cause of suffering and death, 21, 29–31; as a classification, 28–9, 34–6; genetic engineering of, 188n11; interest-based rights, 17–18, 32–3; nutritional needs, 18, 22, 34–6, 41–2, 58, 144; obligate carnivores, 34–5, 141–2, 141(t), 144, 188n9; physiology, 22, 28–9; recommendations, 180–1; terminology, 34. *See also* biological classifications; predators

carnivores, plant-based and alternative foods: about, 21–3, 34–6, 47–9, 180; cellular agriculture, 23, 44–5, 180; for companions, 21–2, 34–6, 60–2, 144, 179–80; cultivated meat, 23, 42–4; food in WRCS, 141–2, 141(t); how to shift from meat to plants, 62; non-sentient animals, 23, 146–7; nutrients vs ingredients, 35; plant-based food, 21–3, 34–6, 49, 141–2, 141(t); scavenged meat, 23, 36–8, 48, 141–2, 141(t), 145–7. *See also* cultivated meat; eggs; pet foods; pet foods, ingredients; scavenged meat; veganism and vegetarianism

carnivores, problem of carnivory: about, 17–18, 21–3, 28–31, 47–9, 180; animal neighbours, 83–4; animals without rights, 45–9; collective responsibilities, 37–8, 41, 44–5, 47–9; interest-based rights, 17–18, 32–3, 44; research needed, 41–2, 48–9, 180; selective

extinction, 8, 23, 33–4, 48, 156, 198n1; sentience, 23, 45–9; vegan carnivores, 34–6
Cataldi, Suzanne Laba, 190n6
cats: big cats as carnivores, 141–2, 141(t); as carnivores, 35; collective responsibilities, 98–9; denizenship, 99–100, 194n16; food for, 37–8, 41–2, 57–8; nutritional needs, 34–6, 41–2, 58, 144; as omnivores, 35–6; plant-based food, 34–6, 144; predator problem, 98–9, 103–4, 106, 189n1, 194n15; selective extinction, 33–4, 194n15; 'strays', 73–6, 80–1, 94. See also companion animals; pet foods
cattle: burger vegans' critiques, 108–13, 116–18, 194n6, 195n14; contaminated food, 69; dairy cows, 24, 44, 197n6. See also animal agriculture; burger vegans; carnivores
cellular agriculture: about, 41–4, 180; collective responsibilities, 41, 44–5, 47–9; defined, 43; food for burger vegans, 111, 113; problem of carnivory, 23, 43–4; recommendations, 179–80; taurine, 41–2. See also cultivated meat
Ceva, Emanuela, 14
Charlton, Anna, 7, 25
chickens: backyard chickens, 38–41, 146, 188n16; as companions, 39–40; problem of carnivory, 38–41; protection from predation, 101; suffering and death, 24, 26–7, 187n4; vertical farms, 128. See also animal agriculture; eggs; food production system

children: analogies to companions, 53–4, 57, 72, 74–6, 93; autonomy of, 63–4, 93, 98, 191nn12–13; children's companions, 65–6; responsibility for, 63–4, 72–3, 93, 101. See also humans
choice and autonomy: about, 60–5, 180; animal neighbours, 92; autonomy, 9, 63–4; companions' food, 60–5; obesity, 60–5, 180, 191n10; recommendations, 180; will rights, 9. See also dignity and respect; relational framework
circuses, 138, 142–3
citizenship of animals, 11–12, 79–81, 105–6, 139–40. See also politics and animal ethics
classifications. See animal neighbours, categories; biological classifications
climate change, 19, 138, 172–5, 177. See also collective societal responsibilities; environment
Cochrane, Alasdair, 9, 27, 118–19, 139, 162
collective societal responsibilities: about, 10, 80–1, 105–6, 172–3, 177, 179–82; climate change, 19, 138, 172–5, 177; companion animals, 66–8, 72–3, 80–1, 180; consumerism, 129–30; dependency relationships, 74, 94–5, 106, 193n5; duty to feed, 72, 94, 172–3, 192n16; food justice, 54–5, 80, 83–4, 186n14; harms to humans, 105–6; harms to thieves, 118–22; individual vs collective, 10; justice, 8–9, 14; key questions, 3–5; predator problem, 98–9, 137; problem of carnivory, 37–8, 41, 44–5, 47–9; relational framework,

105–6; Trafalgar pigeons case, 3–4, 82, 94, 106; vertical agriculture, 126–30; victims of rights violations, 98–9; wild animals, 74, 161, 174–7. *See also* companion animals, societal/political issues; dependency and vulnerability; environment; justice; political philosophy; politics and animal ethics

commensal animals, 84–5, 192n1. *See also* animal neighbours, categories

companion animals: about, 18, 50, 80–1, 179–80; analogy to children, 53–4, 57, 63–5, 72, 165; vs animal thieves, 52; collective responsibilities, 50, 52–3, 66–8, 80–1, 180; interest-based rights, 18, 51–2, 70–7; key questions, 3–5, 50; moral issues, 50, 71–2; as 'natural', 57–60, 180; other people's companions, 50, 65–8, 80–1, 192n3; positive rights, 18, 20, 79, 165; as predators, 152, 181; as property, 51, 67; recommendations, 179–82; relational framework, 11–12, 50, 52–7; shelters for, 53–4; vs wild animals, 51, 73–4, 152. *See also* carnivores, problem of carnivory; dignity and respect; relational framework

companion animals, dependency: about, 52–3, 71–6, 180; vs animal friends, 92–5; and breeding, 53, 71–3, 76; as enduring, 52, 71–6, 79; other people's companions, 65–8, 80–1, 192n3; 'strays', 73–6, 80–1, 94; vulnerability, 53–4, 63–7, 71–6, 80–1; vs wild animals, 73–4. *See also* companion animals, guardians; dependency and vulnerability

companion animals, duty to feed: about, 18, 50–4, 70–7, 80–1; children's companions, 65–6; collective responsibilities, 66–8, 80–1, 180; dependency, 52–4; interest-based rights, 51–4; other people's companions, 65–8, 80–1, 192n3; right to be fed, 18, 70–1, 75–7; state system, 75–7. *See also* companion animals, dependency; companion animals, food; pet foods

companion animals, food: about, 54–65, 179–80; alternatives to plant-based food, 179–80; animal corpses, 36, 38, 57–8, 61; choice in, 60–5, 180; collective responsibilities, 66–8, 80–1, 180; dignity, 55–7; food justice, 54–5, 80, 186n14; 'natural' foods, 57–60, 180; obesity, 14, 60–5, 180, 191n10; plant-based food, 60–2, 144, 179–80; pleasure in, 61–3, 65; recommendations, 179–80; transitions to new foods, 62. *See also* pet foods

companion animals, guardians: choice of food, 60–5; collective responsibilities, 66–70, 80–1; guardians in need, 67–8, 73, 75–6, 192n14; licensing of, 74–5; paternalism of, 63–5; problem of carnivory, 22; vulnerability of companions, 64, 71–3. *See also* collective societal responsibilities; relational framework

companion animals, societal/political issues: about, 11–12, 52, 68–81; analogy to humans, 74–6; citizenship, 79–81, 105–6,

139; collective responsibilities, 66–8, 80–1, 180; food system integration, 68–70, 80; interest-based rights, 51–2, 70–81; laws and regulations, 68–70, 74–5, 80, 106; other people's companions, 65–8, 80–1, 192n3; political inclusion, 75–81; right to be fed, 66–8, 70–7; 'strays', 73–6, 80–1, 94. *See also* companion animals, duty to feed; companion animals, guardians

concrete rights. See *prima facie* and concrete rights

conscious experiences, 189n2. *See also* sentience

consent. *See* choice and autonomy

consequentialism, 8, 10, 117, 161–2

consumer-based solutions, 129–30

contact zone animals, 82–4. *See also* animal neighbours

contaminated food, 68–70. *See also* pet foods

continued life as interest, 26, 33, 161, 178, 199n5. *See also* interest-based rights

Cooke, Steve, 67

corncrakes, 107, 120–1

cows. *See* cattle

Crary, Alice, 190n5

crickets, 46

Cudworth, Erika, 58

cultivated meat: about, 42–4, 49; animal inputs, 42–4, 145–6, 196n5; availability, 44, 189n20; cellular agriculture, 23, 42–4, 48, 179n1, 180n2; collective responsibilities, 44–5, 47–9; human food, 42, 111, 113, 196n5; pet food, 42, 49, 196n5; problem of carnivory, 42–4; WRC food source, 145–7. *See also* carnivores, plant-based and alternative foods; carnivores, problem of carnivory; cellular agriculture

dairy cows, 24, 44, 136, 197n6

Davis, Steven, 110, 113–17, 194n4

Deckers, Jan, 196n5

denizenship, 12, 99–100, 194n16

deontology, 163. *See also* beneficence, duty of; imperfect/perfect duties

dependency and vulnerability: animal neighbours, 83, 87, 92–5; breeding of animals, 53, 71–3, 76; collective responsibilities, 93–5; of companions, 19, 52–4, 63–7, 71–6, 80; internalization of, 71, 74–6; key questions, 178–9; moral duties, 94–5; relational framework, 178–9; 'strays', 73–6, 80–1, 94; Trafalgar pigeons case, 3–4, 82, 94, 106; wild animals, 71, 73–4, 162. *See also* companion animals, dependency; relational framework

Derham, Tristan, 196n1

Despommier, Dickson, 124–5, 128, 195n14

Diamond, Cora, 190n5

dignity and respect: about, 55–7; abolitionism, 7–8, 39–40, 43–4; choice in food, 64–5, 191n13; companion foods, 55–7; corpses, 36, 38, 57–8, 61, 145; dancing bears case, 190n6; 'guardianship principle', 142–3, 147, 198n12; interest-based rights, 190n6; methods of feeding, 56–7; moral individualism, 190n5; moral values, 56–7; wild dignity, 145, 190n3; WRC policies, 154–5. *See also* relational framework

doctrine of double effect, 117–18

dogs: dignity of, 190n3; food preferences, 61; harms vs death, 116; 'natural' foods, 58–60; nutritional needs, 34–6, 58, 144; as omnivores, 34–6, 188n12; plant-based food, 34–6; positive rights, 11; predator problem, 148–9, 189n1, 199n7; 'strays', 73–6, 80–1, 94. *See also* companion animals; pet foods; relational framework

domesticated animals. *See* companion animals

Donaldson, Sue, and Will Kymlicka: citizenship of animals, 12, 79–81, 105, 139–40, 194n16; companions' food preferences, 60–1, 64; cultivated meat, 43; denizenship, 12, 99–100, 194n16; liminal animals, 12, 82–4, 99–100; political philosophy, 12–13, 79; predator problem, 99; sanctuaries, 139–40; sovereignty of wild animals, 12, 199n11; use of animals, 39–40

double effect, doctrine of, 117–18

dumpster diving. *See* scavenged meat

duty to feed. *See* companion animals, duty to feed; interest-based rights; positive rights

eggs: about, 38–41; backyard chickens, 38–41, 48, 188n16; cellular agriculture, 44; collective responsibilities, 41, 44–5, 47–9; food for animals, 38, 40–2, 188n16, 197n6; problem of carnivory, 38–41, 44, 47–9; scale of production, 38, 40–1; suffering and death, 24. *See also* chickens

enemies, animals as. *See* animal neighbours, foes; animal thieves

environment: about, 60; vs animal ethics, 158; bird conservation, 120–1; climate change effects, 19, 138, 172–5, 177; collective responsibilities, 19, 80–1; companions as 'unnatural', 57–60, 191n9; displaced animals, 138; locavorism, 191n8; plant-based food, 60; positive rights of wild animals, 158; renewable energy, 127; rewilded habitats, 125, 132–4, 161, 196n18, 198n1. *See also* collective societal responsibilities; nature and naturalness; WRCS (wildlife rehabilitation centres)

ethics. *See* animal ethics; collective societal responsibilities; food and philosophy; interest-based rights; politics and animal ethics; rights-based approaches

euthanasia, 153, 173–4

extinction, selective, 33–4

farms. *See* agriculture; animal agriculture; arable agriculture; food production system; vertical agriculture

feeding. *See* food and philosophy

feminism, 7–8, 11, 32, 52. *See also* relational framework

field animals. *See* animal thieves; animal thieves, harms; animal thieves, prevention of harms; arable agriculture

Fischer, Bob, 45, 114–15, 120

fish and 'sea food': cellular agriculture, 44; dignity of, 190n3; moral necessity for veganism, 26–7; negative rights, 26–7; pet food, 33; reproduction, 159–61; as rights bearers, 33, 129, 157; sentience, 45–7, 146–7; statistics by weight, 23; suffering and death,

23, 26–7; vertical farms, 128–9. *See also* animal agriculture
fodder, animal food as just, 4–5
foes. *See* animal neighbours, foes
food. *See* agriculture; carnivores, plant-based and alternative foods; companion animals, food; fish and 'sea food'; food production system; meat-eating; pet foods; veganism and vegetarianism; WRCS (wildlife rehabilitation centres), food
food justice, 54–5, 80, 83–4, 186n14. *See also* political philosophy
food and philosophy: about, 13–20; ancient philosophy, 13–14, 185nn12–13, 186n16, 188n9, 190n5; animal ethics, 14–17; animal food as just fodder, 4–5; food ethics, 14–20; key questions, 3–5; moral philosophy, 14–15; philosophy of food, 15–17; political philosophy, 14–15; problem of carnivory, 22, 47–9; relational framework, 86. *See also* animal ethics; carnivores, problem of carnivory; interest-based rights; political philosophy; relational framework; rights-based approaches
food production system: about, 54–5, 68–70; collective responsibilities, 44–5, 80–1; contaminated food, 69–70; food justice, 54–5, 80, 83–4, 186n14; globalization, 69; integration of human and pet systems, 68–70, 80; laws and regulations, 68–70, 80; suffering and death, 23–7. *See also* agriculture; animal agriculture; arable agriculture; cellular agriculture; cultivated meat; eggs; pet foods; suffering and death; vertical agriculture
food studies, 14, 21–2, 68–9, 185n12
foxes, 67, 83, 100–1, 166, 189n1
Francione, Gary, 7, 25
freedom to choose. *See* choice and autonomy
freeganism, 36, 38, 146. *See also* scavenged meat
free-living animals. *See* wild animals
friends. *See* animal neighbours, friends

gardens: animal neighbours in, 82–4; backyard chickens, 39–41, 48, 188n16; burger vegans, 195; collective responsibilities, 10, 41, 95, 123, 126, 181; dependency, 94–5, 181; food sources, 83–4; harms to animals, 122–3; hospitality, 90–2, 95–7, 106, 181; perfect/imperfect duties, 95, 163; predator problem, 96, 100–4; problem of carnivory, 22, 41; recommendations, 83, 181; vertical farms, 122–3. *See also* animal neighbours; birds
Garner, Robert, 184n6
gazelles, 164–6, 199n5
Gillen, Jed, 188n9
Good Samaritans, 67, 89, 92, 95, 106, 193n13. *See also* hospitality
Gruen, Lori, 190n3
guardians for pets. *See* companion animals, guardians

harvesting and harms. *See* animal thieves; animal thieves, harms; animal thieves, prevention of harms; arable agriculture, harms to animals
hedgehogs, 40, 95–6, 100, 146

Heldke, Lisa, 186n17
herbivores, 28–9, 34–6, 141–2, 141(t), 156
Hobbes, Thomas, 183n2
homeless guardians, 67, 73, 192n14. *See also* companion animals, guardians
Horta, Oscar, 160
hospitality: about, 16–18, 89–92, 96–7, 106; acceptance of, 97; analogy to humans, 88–9, 91–3, 96–7; degrees of, 97, 106, 193n10; Good Samaritans, 67, 89, 92, 95, 106, 193n13; imperfect/perfect duties, 89, 90, 96, 163, 193n7; limits on, 90; motivations, 89–90, 104; obligations, 95–7, 101, 106; in philosophy of food, 16–17, 91, 183n2, 193n7, 193n8; safety responsibilities, 95–101, 104, 106; sparrows/sparrowhawk case, 102–4. *See also* animal neighbours
Howard, Len, 93, 97
humans: animal harms to, 100, 105–6; cannibalism, 43, 188n15, 189n19; catastrophic moral horror, 163–4, 169, 175; climate change, 19, 138, 172–5, 177; cultivated meat inputs from, 43–4; displaced animals, 138; food justice, 54–5, 83–4, 186n14; hospitality, 88, 91–3; human diets, 63–5, 191n10, 194n6; and ideologies, 28–9; interest in continued life, 26, 33, 161, 178, 199n5; key questions, 3–5; as *K*-strategist animals, 160–1; 'natural' interactions with animals, 57–60; as omnivores, 28; population growth, 118–19; relational framework, 105–6; speciesism, 6, 56, 79, 164, 199n4; urban wildlife, 82–4; will rights, 9.

See also agriculture; burger vegans; carnivores; children; companion animals, guardians; environment; food production system; meat-eating; omnivores; relational framework; veganism and vegetarianism
hunting by animals. *See* carnivores, problem of carnivory; predators; wild animals; WRCs (wildlife rehabilitation centres); WRCs (wildlife rehabilitation centres), predator problem
hunting by humans: food for burger vegans, 111, 112–13; food for WRCs, 143–4, 147; negative rights, 26–7, 157; for sport, 164
hydroponics, 123, 129. *See also* vertical agriculture

ignored animals: animal neighbors, 84, 88, 97, 103; hands-off approach to wild animals, 51, 157–8, 167–8
imperfect/perfect duties: about, 163, 193n7; beneficence to wild animals, 19, 163–71, 176–7, 199n4; hospitality to neighbours, 89, 90, 96, 193n7
inclusion. *See* relational framework
individual responsibilities: vs collective actions, 10, 80–1. *See also* collective societal responsibilities; companion animals, guardians; companion animals, societal/political issues
indoor agriculture. *See* vertical agriculture
industrial agriculture. *See* agriculture; animal agriculture; arable agriculture; vertical agriculture
insects: as food, 49, 113;

insectivores, 28, 29; problem of carnivory, 45–7; sentience, 45–7, 160, 180. *See also* invertebrates
institutional responsibilities. *See* collective societal responsibilities; companion animals, societal/political issues
interest-based rights: about, 8–13, 32–3, 70–5, 178; companion animals, 51–2, 70–5; continued life, 26, 33, 161, 178, 199n5; individual vs collective responsibilities, 10, 72, 80–1; justice issues, 70–2; key questions, 3–5; as mainstream framework, 185n10; negative rights, 10–12, 26–7, 32–3, 51, 70–1, 75, 157, 178–9; negative vs positive rights, 32–3; positive rights, 11, 32–3, 72, 79, 178–9; problem of carnivory, 26, 32–3, 47–9; relational framework, 11–12; rights-based approaches, 32–3, 51–2; sentience, 9–10, 45, 70, 72, 178, 189n2; suffering and death, 9–10, 26–7, 72; and utilitarianism, 10; vs will rights, 9. *See also* collective societal responsibilities; *prima facie* and concrete rights; relational framework; sentience; suffering and death
interspecies friendship, 86–7. *See also* animal neighbours, friends; companion animals; relational framework
invertebrates: about, 46; as animal thieves, 107; farming of, 129, 146–7, 195n15; food for birds, 46, 83, 100; food for burger vegans, 111, 113; food for carnivores, 46, 146–7; food for companions, 180; food from cellular agriculture, 44; harms from arable agriculture, 115; insects, 28, 29, 45–7, 49, 113, 160; problem of carnivory, 44, 46–7, 49; recommendations, 179–80; sentience problem, 45–7, 107, 146–7, 160, 180; worms, 46, 83, 100, 129, 195n15
Irvine, Lesley, 67, 192n14

Johannsen, Kyle, 19, 163–71
just fodder, animal food as, 4–5
justice: about, 8–10; duty of justice vs beneficence, 158, 165; as duty of the state, 171; enforcement of rights, 51; meat-eating, 26; natural vs just world, 139, 156; normative obligations, 51, 178. *See also* collective societal responsibilities; companion animals, societal/political issues; interest-based rights; politics and animal ethics

Kaiser, Matthias, 14, 185n12
Kant, Immanuel, 91, 183n2
Kaplan, David, 14, 15, 185n12
Keulartz, Jozef, 196n2
Knight, Andrew, 144
K-strategist animals, 160–1
Kymlicka, Will. *See* Donaldson, Sue, and Will Kymlicka

Lamey, Andy, 114, 115, 120, 194n1
legal systems. *See* collective societal responsibilities; companion animals, societal/political issues
Leitsberger, Madelaine, 144
liminal animals, 12, 82–4, 99–100, 134. *See also* animal neighbours
lions, 37, 138, 142–3, 164–6, 199n6
lizards, 65–6, 99
Lomasky, Loren, 187n8

marine animals. *See* fish and 'sea food'
Marra, Peter P., 99, 194n15
Mathews, Freya, 196n1
McMahan, Jeff, 156
meat-eating: about, 22–3; by burger vegans, 107–8, 110, 112, 116; cannibalism, 43, 188n15, 189n19; cultural meanings, 25, 27, 30, 61–2; ethics of, 16, 22–3; greenwashing/humanewashing meat, 146; justice issues, 26; omnivores, 28; problem of carnivory, 22–3, 47–9. *See also* animal agriculture; carnivores; carnivores, problem of carnivory; scavenged meat; veganism and vegetarianism
melamine contamination, 69
mice, 84, 107, 114, 116–18, 131–2, 197n7
milk production, 24, 44, 83, 146, 197n6
Mill, John Stuart, 184n3
Monroe, Dave, 185n12
moral agency: agents vs non-agents, 71–2, 98–9, 103–4, 106, 148; dilemma in *Star Trek*, 149–53, 197nn8–10; rights violations by agents, 103–4, 148, 164–6
moral philosophy: about, 5–8, 12–13; care ethics, 7–8, 32, 52; catastrophic moral horror, 163–4, 169, 175; interest-based rights, 10, 51; moral individualism, 190n5; vs political philosophy, 51. *See also* animal ethics; deontology; food and philosophy; hospitality; imperfect/perfect duties; interest-based rights; politics and animal ethics; relational framework; utilitarianism

mutualistic relations, 192n1

Nagel, Thomas, 189n2
Narveson, Jan, 187n8
nature and naturalness: companions' foods, 57–60; food at WRCs, 135–6, 140–2, 141(t), 147; laissez-faire approach, 157–8; natural vs just world, 139, 156; suffering and death, 157. *See also* environment
negative rights: about, 5, 10–12, 26–7, 77, 178; examples, 10–11; interest-based rights, 10–12; vs positive rights, 32–3, 75, 199n9; relational framework, 12. See also *prima facie* and concrete rights; suffering and death
neighbours, animal. *See* animal neighbours
Nesheim, Malden C., 21–2, 35, 68–70
Nestle, Marion, 21–2, 35, 68–70, 185n12
non-domesticated animals in human environments. *See* animal neighbours; animal refugees; animal thieves; sanctuaries; WRCs (wildlife rehabilitation centres)
non-domesticated animals in non-human environments. *See* wild animals
nonhuman animals, terminology, 183n1. *See also* animals
non-sentient animals, 146–7. *See also* invertebrates; sentience
normative obligations: about, 5–6, 14–17; act of eating, 86; animal thieves, 18–19; companion animals, 18; food ethics, 14–17; food studies, 14, 21–2, 68–9, 185n12; key questions, 3–6; as political, 11–12, 51; relational

framework, 11–12, 86. *See also* animal ethics; beneficence, duty of; collective societal responsibilities; imperfect/perfect duties; relational framework
Nussbaum, Martha, 56, 139, 156, 190n5

obesity, 14, 63–5, 180, 191n10
Okuleye, Yewande, 184n6
omnivores: as a classification, 28–9, 34–6; continuum of, 34–5; dogs as, 34, 188n12; plant-based food, 34–6; problem of carnivory, 22–3, 28–33, 34. *See also* biological classifications; carnivores, problem of carnivory
Oxford Vegetarians, 6, 184n5

pain. *See* suffering and death
Palmer, Clare, 12–13, 52–3, 82–4, 94, 106, 138, 157–8
parasitic relations, 192n1, 192n4. *See also* animal neighbours
paternalism, 63–5. *See also* choice and autonomy; companion animals, guardians
perfect duty. *See* imperfect/perfect duties
pet foods: about, 21–2, 49, 68–70; access to information on, 55, 59, 69–70, 77; collective responsibilities, 80–1; for companions, 41–2, 54; food banks, 76; food justice, 54–5, 80, 83–4, 186n14; integration with human food system, 68–70, 80; laws and regulations, 68–70, 80; 'natural' foods, 57–60; pleasure in eating, 42, 61–2; preferences, 60–5; problem of carnivory, 21–2; recommendations, 83; research needed, 41–2, 48–9, 180; terminology, 187n1; transitions to new foods, 62. *See also* carnivores, plant-based and alternative foods; carnivores, problem of carnivory; companion animals, duty to feed; companion animals, food
pet foods, ingredients: about, 21–2; contaminated food, 59, 68–70; fish, 33; invertebrates, 46, 49; nutrition, 34–6, 41–2, 58–9; plant-based food, 22–3, 49, 61–2; scavenged meat, 37–8, 49; taurine supplements, 41–2
pets. *See* companion animals
philosophy and food. *See* food and philosophy
Pierce, Jessica, 61–2, 64, 188n9
pigeons, 3–4, 82, 84, 94, 99, 106, 146
pigs, 45, 69, 112, 128, 194n6
pigs, wild, 141–2, 141(t)
plant-based food. *See* carnivores, plant-based and alternative foods; veganism and vegetarianism
Plato, 13, 185nn12–13, 186n16, 188n9
polar bears, 173–5, 200nn12–13
political philosophy: about, 5–6, 12–13, 14, 51; food ethics, 14–15; food justice, 54–5, 83–4, 186n14; institutional issues, 14; interest-based rights, 8–10, 51; justice issues, 14; vs moral philosophy, 51; normative obligations, 11–12; rights issues, 51, 105; value-based rights, 8–9. *See also* collective societal responsibilities; food and philosophy; interest-based rights; justice
politics and animal ethics: about, 8–13; citizenship, 79–81, 105, 139–40; harms to humans,

105–6; individual vs collective responsibilities, 10–12, 80–1, 105–6, 126–8; political inclusion, 75–81; problem of carnivory, 37–8, 41, 47–9; vertical agriculture, 126. *See also* collective societal responsibilities; companion animals, societal/political issues; interest-based rights; justice

population control, 94, 118–19, 121, 130, 161

pork, 45, 69, 112, 128, 194n6

Porphyry, 184n4

positive rights: about, 5, 10–11, 20, 77, 79; examples of, 5, 11, 178–9; interest-based rights, 10–11; vs laissez-faire approach, 157–8; vs negative rights, 32–3, 75, 199n9; relational framework, 11, 20. *See also* interest-based rights; relational framework

Posner, Richard, 193n12

poultry. *See* chickens

precautionary principle, 45–6

predators: about, 98–102, 147–55, 164–71, 180–1; analogy to humans, 98, 101, 151–2; analogy to *Star Trek* dilemma, 149–51, 153, 197nn8–10; collective responsibilities, 98–9, 103–4, 148–51, 157, 180–1, 196n2; denial of protection for foes, 101–4; duty to protect companions, 189n1, 199n7; duty to protect friends, 98–101; lack of moral agency, 150, 164–6; negative/positive rights, 148–51; obligate carnivores as non-predatory, 34–5, 141–2, 141(t), 144, 147, 188n9; predator problem, 56, 98–102; recommendations, 180–1; rewilded habitats, 161, 196n18; risks to animal neighbours, 95–100; selective extinction, 8, 23, 33–4, 48, 156, 198n1; sparrows/sparrowhawk case, 102–4; suffering of prey, 156–7. *See also* carnivores, problem of carnivory; wild animals; WRCS (wildlife rehabilitation centres), predator problem

prima facie and concrete rights: concrete rights, 26–7, 162; interest-based rights, 32–3; negative rights, 9–10, 26, 32–3; *prima facie* rights, 26–7, 32–3, 36, 101, 158, 162; protection from predators, 101; wild animals in need, 158–62

property, animals as, 51, 67

property/territorial rights, 85, 168–71, 199n8, 199nn10–11. *See also* wild animals

proportionality, 162, 166, 199n5

rabbits, 67, 95, 101, 141–2, 141(t), 146, 150–1

raccoons, 3, 11, 84, 193n12

Rachels, James, 190n5

rats, 18, 85, 107, 187n2. *See also* animal neighbours; animal thieves

refugees. *See* animal refugees

Regan, Tom: burger vegans, 110, 114, 116, 194n4; harms vs death, 116; rights approach, 114, 142, 148, 184n7, 184n9, 195n7; value-based rights, 7–9; WRC predator problem, 198nn12–13

rehabilitation, as concept, 140–2, 141(t). *See also* sanctuaries; WRCS (wildlife rehabilitation centres)

relational framework: about, 20, 52–4, 162–3, 178–9; affective relations, 52, 67–8, 166, 179; belonging, 179; care ethics, 7–8, 32, 52; companions, 39–40, 52–4, 67–8, 166; dependency, 52–4, 178–9; food and eating, 86, 89–90; harms to humans, 100, 105–6; hospitality, 90; human-human relations, 67–8; interest-based rights, 11, 52; key questions, 3–5, 178–9; normative obligations, 11–12, 86; wild animals, 151–2, 159, 162–3, 190n3, 198n3. *See also* choice and autonomy; collective societal responsibilities; companion animals, dependency; dependency and vulnerability; dignity and respect; hospitality

reproduction, 159–61

respect. *See* dignity and respect; relational framework

rewilded habitats, 125, 132–4, 161, 196n18

rights, animals without, 45–9. *See also* sentience

rights-based approaches: about, 7–8, 51, 162–3; animals without rights, 44–9; catastrophic moral horror, 163–4, 169, 175; critiques of burger vegans, 116–17; duty of beneficence, 165, 171; individual vs collective responsibilities, 80–1; interest- vs value-based rights, 8–9; as justice issues, 26–7; least harms, 116–17; moral individualism, 190n5; negative rights, 26–7, 178; negative vs positive rights, 32–3, 75, 199n9; *prima facie* and concrete rights, 26–7, 162; relational framework, 162, 178; sentience, 45–9. *See also* collective societal responsibilities; food and philosophy; imperfect/perfect duties; interest-based rights; *prima facie* and concrete rights

roadkill, 36–8, 48, 111, 113, 115, 145. *See also* scavenged meat

rodents, 84, 107, 109, 181n6. *See also* mice

Rodman, John, 58
Rolston, Holmes, III, 58
Rothgerber, Hank, 21
Rousseau, Jean-Jacques, 184n3
Royal Society for the Protection of Birds (RSPB), 120, 195n11
Ryder, Richard, 184n4

Salt, Henry S., 184n4
sanctuaries: about, 135–7; for birds, 94; for farmed animals, 139–40, 153–5; food sources, 40; problem of carnivory, 22, 40, 42; reimagined communities, 139–40, 153–5; release policies, 153–5. *See also* carnivores, problem of carnivory; WRCs (wildlife rehabilitation centres); WRCs (wildlife rehabilitation centres), predator problem

Santella, Chris, 99, 194n15

scavenged meat: about, 36–8, 48; for burger vegans, 111, 113, 115; collective responsibilities, 37–8, 41, 44–5, 47–9; legal questions, 188n14; obligate carnivores, 141–2, 141(t); problem of carnivory, 23, 36–8, 44, 47–9; roadkill, 37–8, 48, 111, 113, 115, 145; waste products, 48; WRC foods, 141–2, 141(t), 145–7. *See also* carnivores, plant-based and alternative foods; carnivores, problem of carnivory

sea food. *See* fish and 'sea food'

selective extinction, 8, 23, 33–4, 48, 156, 198n1

sentience: about, 9–10, 45–7, 146–7, 160; animals without rights, 45–9; collective responsibilities, 47–9; companion animals, 70, 72; defined, 7, 189n2; edge cases, 45–9; farming of non-sentient animals, 146–7; interest-based

rights, 9–10, 26, 70, 72, 189n2; interest in continued life, 26, 33, 150, 161, 178, 199n5; invertebrates, 45–7; *K*- and *r*-strategist animals, 160–1; negative rights, 26–7, 32–3, 178; precautionary principle, 45–7; problem of carnivory, 23, 45–9; problem of non-sentience, 47–8, 146–7, 160, 180; utilitarian approach, 7–8
sheep, 39–40, 45, 69, 195n14
shelters for companions, 53–4
shelters for non-domesticated animals. *See* sanctuaries; WRCS (wildlife rehabilitation centres)
Singer, Peter, 6–10, 45–6, 117, 161–2, 184nn6–7, 190n5. *See also* utilitarianism
societal responsibilities. *See* collective societal responsibilities
Socrates, 185n12, 188n9
sovereignty of wild animals, 12, 85, 168–72, 176–7, 199n11
sparrowhawks, 102–4, 159, 194n17
sparrows, 11, 82, 87, 100, 102–4
speciesism, 6, 56, 79, 164, 199n4
Star Trek, moral dilemma, 149–53, 197nn8–10
state responsibilities. *See* collective societal responsibilities
stoats (weasels), 150–1, 198n12
strangers, animal. *See* wild animals
suffering and death: about, 23–7, 159–64, 182; catastrophic moral horror, 163–4, 169, 175; climate change effects, 19, 138, 172–5, 177; duty of beneficence, 19, 158, 163–71, 176–7, 199n4; euthanasia, 153, 173–4; imperfect duties, 163–4; interest-based rights, 9–10, 26; interest in continued life, 26, 33, 150, 161, 178, 199n5; negative rights, 10–11, 26–7; problem of carnivory, 21, 29–33; recommendations, 182; reproduction, 159–61; sentience, defined, 7; statistics, 23; utilitarian critique, 6; wild animals, 157, 182, 199n5. *See also* interest-based rights; negative rights; sentience
Szymanski, Ileana, 86

taurine, 41–2
Taylor, Angus, 194n4
Telfer, Elizabeth, 16, 89–92, 95–6, 193nn7–10
territorial/property/sovereignty rights of wild animals, 12, 85, 168–72, 176–7, 199n8, 199nn10–11. *See also* wild animals
testing, animal, 129, 142, 170
thieves. *See* animal thieves
Thompson, Paul B., 185n12
Tomkins, Bruce, 127
Townley, Cynthia, 87
Trafalgar Square pigeons, 3–4, 82, 94, 106

utilitarianism: about, 6–10; consequentialism, 8, 10, 117, 161–2; feminist critiques, 7–8; interest-based rights, 10; vs perfect duties, 163

Varner, Gary, 189n2
veganism and vegetarianism: about, 3–4, 23–7, 181; burger vegans' critiques of, 107–13; vs carnism, 28, 187n6; carnivore food, 34–6; cultural meanings, 25, 27; justice issues, 26–7; key questions, 3–5; moral necessity of, 7, 23–7; practices vs ideologies, 28–9, 187n6; problem of

carnivory, 22–3, 28–36, 47–9; recommendations, 181; suffering and death of animals, 23–7; veganism vs vegetarianism, 187n3. *See also* agriculture; burger vegans; carnivores; carnivores, plant-based and alternative foods; carnivores, problem of carnivory; food production system

vermiponics, 129, 195n15. *See also* vertical agriculture

vertical agriculture: about, 18–19, 122–34, 181; animal agriculture, 128–30, 195n14; animal thieves, 125–6, 131–2; collective responsibilities, 126–30, 181; crops, 124, 128; diseases, 125–6; harms to animals, 108, 122–3, 125–6, 130–4; loss of habitat, 131–2; 'ponic' farms, 123, 128, 129, 195n15; productivity, 124–5, 127–8; recommendations, 181; as response to burger vegans, 108, 126, 133–4; rewilded habitats, 125, 132–4, 196n18; small scale, 122–3; structures, 123–5, 127–31, 195n13. *See also* animal thieves, prevention of harms

virtues. *See* hospitality; imperfect/perfect duties; nature and naturalness

Voyager (Star Trek), moral dilemma, 149–53, 197nn8–10

vulnerability. *See* dependency and vulnerability

vultures, 141–2, 141(t). *See also* scavenged meat

Wayne, Katherine, 33–4
weasels (stoats), 150–1, 198n12
Wild Animal Initiative, 156
wild animals: about, 19, 137–9, 156–71, 182; analogy to *Star Trek* dilemma, 149–53, 197nn8–10; animal vs environmental ethics, 158; catastrophic moral horror, 163–4, 169, 175; climate change effects, 19, 138, 172–5, 177; collective responsibilities, 12, 137–40, 152, 161–2, 167–8, 170–7, 193n12; vs companions, 51, 73–4, 152; dependency, 71, 73–4, 162; dignity, 145, 190n3; duty of beneficence, 19, 158, 163–71, 176–7, 199n4; hands-off approach, 51, 157–8, 167–8; imperfect duties, 163–5, 167, 175–7; key questions, 3–5; moral agency, 164–6; negative rights, 5, 148–51, 153, 157, 159, 199n9; positive obligations, 5, 139–40, 148, 152–3, 156–9, 162–3, 199n9; practicality of, 158, 165; predators, protection from, 164–72, 176; *prima facie* rights, 26–7, 158–64; property/territorial/sovereignty rights, 12, 85, 168–72, 176–7, 199n8, 199nn10–11; proportionality, 162, 166, 199n5; recommendations, 181–2; relational framework, 151–2, 159, 162–3, 190n3, 198n3; reproduction, 159–61; vs 'strays,' 73–6, 80–1, 94; suffering and death, 26–7, 156–7, 159–61, 169, 173–7, 182. *See also* animal refugees; animal thieves; predators; sanctuaries; WRCs (wildlife rehabilitation centres)

wild animals, specific interventions: about, 161–2, 167–9, 182; euthanasia, 153, 173–4; food, 161–2, 171–7, 181–2; genetic editing, 161, 167–70; inoculations, 161–2; polar bears case, 173–5,

200nn12–13; population control, 161–2, 167; relocations, 173–4; restructuring wild spaces, 161–2, 168–9, 176–7; rewilded habitats, 125, 132–4, 161, 196n18

will rights, 9. *See also* interest-based rights

Winters, Ed, 188n16

Wittgenstein, Ludwig, 190n5

worms, 46, 83, 100, 129, 195n15

WRCs (wildlife rehabilitation centres): about, 19, 137–42, 141(t), 153–5, 181; analogy to *Star Trek* dilemma, 149–53, 197nn8–10; animals as refugees, 136–40, 142–5; collective responsibilities, 137–8, 143, 151–5; negative rights, 153; positive rights, 139–40, 156–8; recommendations, 181; rehabilitation, as concept, 140–2; reimagined communities, 139–40, 153–5; release policies, 136, 141, 147–8, 152–5. *See also* animal refugees

WRCs (wildlife rehabilitation centres), food: about, 135–7, 141–7, 142(t), 154–5, 181; cultivated meat, 145–7; cultivated milk, 146, 197n6; difficult vs easy to feed, 141–2, 141(t), 147; food sources, 135–7, 142–7, 154–5, 197n6, 198n12; 'guardianship principle', 142–3, 147, 198n12; hunted animals, 142–4, 147; 'natural' foods, 135–6, 144–5; plant-based food, 144, 147, 197n6; rescued animals as, 135–6, 145–6, 147; scavenged meat, 141–2, 141(t), 145–7. *See also* carnivores, plant-based and alternative foods; carnivores, problem of carnivory

WRCs (wildlife rehabilitation centres), predator problem: about, 19, 137, 147–55, 181–2; analogy to *Star Trek* dilemma, 149–53, 197nn8–10; collective responsibilities, 19, 137, 148–55, 181–2, 196n2, 197n7; euthanasia, 153; hands-off approach, 148, 157–8; predatory vs non-predatory, 141–2, 141(t), 147; recommendations, 137, 153–5, 181–2; refusal to accept predators, 153; refusal to release, 137, 152–5; reimagined communities, 139–40, 153–5. *See also* carnivores, problem of carnivory

Wrenn, Corey Lee, 137–8, 142–7, 154, 196n5

zoos, 56, 138, 161, 167

Zwart, Hub, 14–15